SpringerBriefs in Petroleum Geoscience & Engineering

Series Editors

Dorrik Stow, Institute of Petroleum Engineering, Heriot-Watt University, Edinburgh, UK

Mark Bentley, AGR TRACS International Ltd, Aberdeen, UK

Jebraeel Gholinezhad, School of Engineering, University of Portsmouth, Portsmouth, UK

Lateef Akanji, Petroleum Engineering, University of Aberdeen, Aberdeen, UK

Khalik Mohamad Sabil, School of Energy, Geoscience, Infrastructure and Society, Heriot-Watt University, Edinburgh, UK

Susan Agar, Oil & Energy, Aramco Research Center, Houston, USA

Kenichi Soga, Department of Civil and Environmental Engineering, University of California, Berkeley, USA

A. A. Sulaimon, Department of Petroleum Engineering, Universiti Teknologi PETRONAS, Seri Iskandar, Malaysia

T0172256

The SpringerBriefs series in Petroleum Geoscience & Engineering promotes and expedites the dissemination of substantive new research results, state-of-the-art subject reviews and tutorial overviews in the field of petroleum exploration, petroleum engineering and production technology. The subject focus is on upstream exploration and production, subsurface geoscience and engineering. These concise summaries (50-125 pages) will include cutting-edge research, analytical methods, advanced modelling techniques and practical applications. Coverage will extend to all theoretical and applied aspects of the field, including traditional drilling, shale-gas fracking, deepwater sedimentology, seismic exploration, pore-flow modelling and petroleum economics. Topics include but are not limited to:

- Petroleum Geology & Geophysics
- Exploration: Conventional and Unconventional
- Seismic Interpretation
- Formation Evaluation (well logging)
- Drilling and Completion
- Hydraulic Fracturing
- Geomechanics
- Reservoir Simulation and Modelling
- Flow in Porous Media: from nano- to field-scale
- Reservoir Engineering
- Production Engineering
- Well Engineering; Design, Decommissioning and Abandonment
- Petroleum Systems; Instrumentation and Control
- Flow Assurance, Mineral Scale & Hydrates
- Reservoir and Well Intervention
- Reservoir Stimulation
- Oilfield Chemistry
- Risk and Uncertainty
- Petroleum Economics and Energy Policy

Contributions to the series can be made by submitting a proposal to the responsible Springer contact, Charlotte Cross at charlotte.cross@springer.com or the Academic Series Editor, Prof Dorrik Stow at dorrik.stow@pet.hw.ac.uk.

More information about this series at http://www.springer.com/series/15391

Qiuyang Shen · Jiefu Chen · Xuqing Wu ·
Yueqin Huang · Zhu Han

Statistical Inversion of Electromagnetic Logging Data

Springer

Qiuyang Shen
Cyentech Consulting LLC
Cypress, TX, USA

Jiefu Chen
Electrical and Computer Engineering
University of Houston
Houston, TX, USA

Xuqing Wu
Information and Logistics Technology
University of Houston
Houston, TX, USA

Yueqin Huang
Cyentech Consulting LLC
Cypress, TX, USA

Zhu Han
Electrical and Computer Engineering
University of Houston
Houston, TX, USA

ISSN 2509-3126   ISSN 2509-3134   (electronic)
SpringerBriefs in Petroleum Geoscience & Engineering
ISBN 978-3-030-57096-5   ISBN 978-3-030-57097-2   (eBook)
https://doi.org/10.1007/978-3-030-57097-2

This Springer imprint is published by the registered company Springer Nature Switzerland AG
The registered company address is: Gewerbestrasse 11, 6330 Cham, Switzerland

Contents

Acronyms

BHA	Bottom Hole Assembly
DoI	Depth of Investigation
EM	Electromagnetic
HAHZ	High Angle and Horizontal
HMC	Hybrid/Hamiltonian Monte Carlo
LMA	Levenberg-Marquardt Algorithm
LWD	Logging-While-Drilling
MCMC	Markov Chain Monte Carlo
MHA	Metropolis-Hastings algorithm
PDF	Probability Density Function
PPD	Posterior Probability Distribution
tMCMC	Trans-dimensional Markov Chain Monte Carlo
TVD	True Vertical Depth

Chapter 1
Introduction

Abstract It is an art to learn the story beneath our feet. Human used to travel hundreds of thousands of miles and accomplished a space trip to the moon, whereas the journey into the earth can never penetrate even the shell. For centuries, geoscientists have been practicing new technologies adopted from inter-disciplines of mechanics, physics, chemistry, and mathematics to improve the understanding of inside earth. As a planetary science, geology concerns with the earth composition and its change over time. Figuring out the geological structure is essential for mineral and hydrocarbon exploration. Drilling, as a practice in the exploration, helps people learn the geological properties through direct contact with earth formation. In this chapter, we will introduce the fundamental concepts of drilling and well logging, electromagnetic logging-while-drilling service, and the mathematical tools people used to extract the information hidden behind the drilling data. Starting from it, we will reveal the topic of this book, a powerful tool for logging data interpretation, statistical Bayesian inference.

1.1 Background of Well Logging

Unlike the geophysical study of the subsurface environment, which is often conducted on the surface and tomographying the earth structure underneath, the well logging is a drilling practice that happens inside the earth. Multiple records are collected while borehole penetrates the geological formations. A group of sensors are equipped on the downhole logging instrument and taking physical measurements of the surrounding formation. Doctors can tell a potential heart problem from electrocardiogram. Similarly, geoscientists analyze and interpret the well logs and reveal the hidden structure, more specifically, the physical properties of earth formation.

Well logs usually consist of multiple types of measurements which are rendered by different sensing systems. The petrophysical properties in or around a well can be determined based on the interpretation of well logs (Fig. 1.1). For example, the lithology logs generated by the gamma ray logging tools tell the natural radioactivity of formation. It helps distinguish sands from shales. Porosity logs measured

© The Author(s), under exclusive license to Springer Nature Switzerland AG 2021
Q. Shen et al., *Statistical Inversion of Electromagnetic Logging Data*,
SpringerBriefs in Petroleum Geoscience & Engineering,
https://doi.org/10.1007/978-3-030-57097-2_1

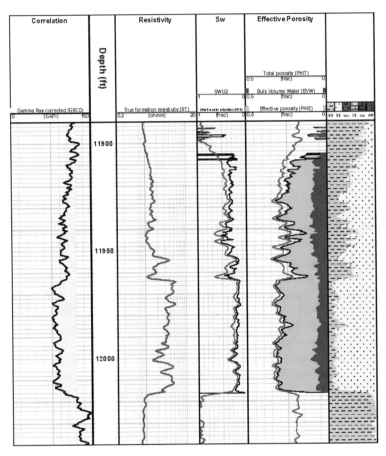

Fig. 1.1 An example of wireline logs consisting of multiple measurements versus depth. The rock composition can be derived from the petrophysical properties and is shown in the right column. Author's right: Andy May

by acoustic or nuclear equipment reflect the percentage of pore volume in rocks. Among others, the resistivity logs that measure the electrical properties are capable of differentiating the conductive salt water and resistive hydrocarbons.

The exercises of well logging are mainly executed through two different drilling operations, the wireline, and logging-while-drilling (LWD). The former one conducts well logging by lowering measurement devices into an open well using an electrical cable. The measured data are transmitted back to the ground via cable. LWD, in its literal sense, conducts well logging while during drilling process. It makes the formation evaluation on the fly. The sensor systems are integrated into the drilling string as a bottom hole assembly (BHA) and taking the real-time measurements. The measured data are transmitted partially to the surface via telemetry system (usually the mud pulse or electromagnetic telemetry), while a complete dataset is stored in the tool embedded memory chips. An advanced LWD operation can make many tough tasks possible. Landing to the target zone and navigating in an optimal well path relies on the real-time log interpretation and formation evaluation. This topic is continued in depth and one particular LWD technology is introduced in the next section.

1.2 Directional Electromagnetic LWD Services and Tools

Among multiple LWD techniques, directional electromagnetic (EM) LWD has been promoted and used in high angle and horizontal (HAHZ) drilling applications.

1.2.1 Directional EM LWD Services

The first EM resistivity LWD service was introduced in the 1980s [1, 2] that equipped with coaxial transmitters and receivers. The EM wave propagates through the medium around the drilling borehole, which derives the electrical conductivity of the earth formation. A design of multi-spacing and multi-frequency with compensation was lately standardized as the conventional EM LWD resistivity tools in the 1990s, which provided multiple depth-of-investigation (DoI) capability and met the requirement of more challenging LWD environment [3]. The inadequacy of conventional LWD resistivity tools was the lack of azimuthal sensitivity that provides the directionality information [4]. It made the evaluation of complex reservoir difficult. A revolution of EM LWD tools was made after 2000 and the directional EM LWD tools were developed and field-tested [5–7]. The directional tools equipped with nonaxial (tilted or transversal) antennas which provide an azimuthal sensitivity. As a result, the measurements are dependent on the tool azimuthal orientation while rotating. The new design makes a variety of directional drilling applications achievable since the measurements can indicate the locations of upper and lower boundaries as well as the formation anisotropy. The directionality introduced benefits the accurate formation evaluation, especially to the real-time geosteering. It is an essential technique for proactive guidance in the horizontal wells, which maintains the maximal reservoir contact and avoids boundary breakthrough [8].

There has been a steady advancement in the tool features and capacity [4, 9] until a new generation of directional EM LWD tools were commercialized in the mid 2010s [10]. The service is named after the 'ultra-deep' that brought a great extension of DoI. With the lower frequency and longer T-R spacing, a radial DoI has been extended to over 100 ft (30 m) from the drilling borehole [11]. The geological features in reservoir-scale are within the tool scope. Moreover, with the success of launching ultra-deep service, the applications of the new service are widely extended to well placement, reservoir mapping, geo-stopping, landing, or fault detection. This new service has been under the spotlight and drawn great attention from oil operators. However, due to the complexity of the measurement physics, the tool response characteristics and the data processing, the operators do not have sufficient confidence to use the service as much as expected without a sufficient understanding of the uncertainty of the service. To promote this new technology and service, the Standardization of Deep Azimuthal Resistivity (SDAR) workgroup, which is affiliated to SPWLA Resistivity Special Interest Group (RtSIG) chapter, was formed in 2016 to help the industry gain such understanding in utilizing the ultra-deep directional EM LWD service.

1.2.2 Directional EM LWD Tools

A key role in the service aforementioned is the drilling tool named azimuthal resistivity LWD tool [6, 7].

Figure 1.2 demonstrates a schematic diagram of azimuthal resistivity LWD tool, which is equipped with multiple antennas whose polarization can be along the tool axis (denoted as the Z direction hereinafter), or perpendicular to the axis (X or Y direction), or in a tilted direction. In a most general case, both the transmitter and the receiver are triaxial antennas as shown in Fig. 1.3, and the responses of this pair of transmitter and receiver can be written in a 3-by-3 tensor [12]. Different types of measurements can be obtained from the tensor, which can provide different sensitivities to the formation properties. For example, the ZZ measurements are obtained from the transmitter in Z direction and the receiver in Z direction. This type of measurements provide a standard sensitivity to the formation resistivity. The tools are usually designed in fully compensated, i.e, multiple ZZ measurements are used to reduce the unwanted environmental and temperature variations. Hence, we can derive two further products, or called curves, based on the ZZ measurements from two receivers R1 and R2 as,

$$Att = 20 \times log_{10} \frac{abs(V_{R1})}{abs(V_{R2})}$$
$$Ps = angle(V_{R2}) - angle(V_{R1}),$$

(1.1)

where Att and Ps stand for attenuation and phase difference, respectively. And V denotes the induced voltage in receiver. These type of curves have no directionality,

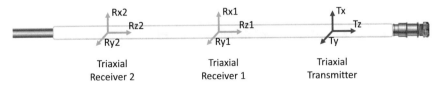

Fig. 1.2 Schematic diagram of an azimuthal resistivity tool. The tool is configured with triaxial transmitter and receiver antennas. The EM waves are transmitted and propagated in the surrounding formation and captured by the receivers

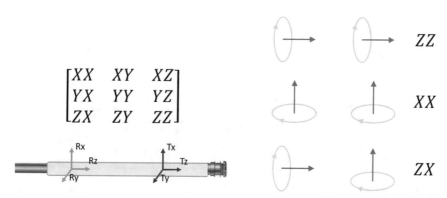

Fig. 1.3 The left panel shows a 9-component tensor obtained by a pair of triaxial transmitter and receiver. The right panel presents a schematic of ZZ, XX, and ZX measurements

i.e. they cannot distinguish up from down, or left from right. They are used for standard measurements of formation resistivity.

Two other measurements, ZX and XX are commonly used in the azimuthal resistivity tools, which can provide tools a directional sensitivity. The ZX measurements are obtained by transmitter with component along the X direction, and receiver in Z direction. Each data point of a ZX measurement is calculated after a full 360° of tool rotation w.r.t. the Z axis. ZX measurements have azimuthal sensitivity, which means they can be used to figure out the azimuthal angle of a bed interface relative to the tool. This property of the ZX curves makes them very suitable for figuring out the azimuth and distance of adjacent boundaries and keep the tool in the desired bed, or in one word, geosteering. The XX measurements are obtained by transmitter with component along the X direction, and X direction receiver. The construction of a data point of a XX measurements also requires a full tool rotation around the Z axis. The XX measurements are sensitive to formation resistivity anisotropy at any relative dip angle, thus they are the best candidates for anisotropy inversion [13].

Like in Eq. 1.1, one can derive further attenuation and phase difference curves from ZX, XX, and other components. Or one can use the combination of multiple components and derive the curves with multiple sensitivities. We derived four types of curves based on the industrial practice as [6, 10, 12] introduced. These four different types of curves are listed below. Each type of curves is sensitive to different earth formation features.

- Symmetrized directional curves: sensitive to the formation boundary.
- Anti-symmetrized directional curves: sensitive to the dip and anisotropy.
- Harmonic resistivity curves: sensitive to the bulk resistivity.
- Harmonic anisotropy curves: sensitive to the resistivity and anisotropy.

For a better understanding of such a tool's response and behavior regarding earth physics, we use a set of sensitivity studies to demonstrate the curve behavior when a synthetic tool is moving on a simple two-layer model as in Fig. 1.4. A synthetic tool

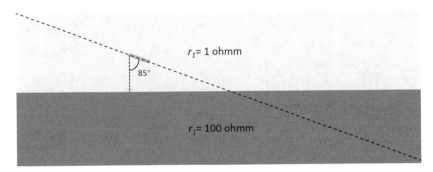

Fig. 1.4 A simple two-layer model with an interface. The tool is moving along the dash trajectory and recording the measurements. A dip angle denotes that the tool axis and the normal line of the interface is 85°

is simulated by one pair of triaxial transmitter and receiver working at a frequency of 12 kHz with T-R spacing of 45 ft.

In the model shown in Fig. 1.4, the upper layer is a conductive isotropic formation whose resistivity is 1 ohm-m. The lower layer is a resistive isotropy formation whose resistivity is 100 ohm-m. The interface between the upper layer and the lower layer is located at 0 ft in true vertical depth (TVD). The dip angle between the tool axis and the normal line of the interface is 85°.

The sensitivity here is defined as the derivative of the tool responses regarding the formation parameters i.e. the bed boundary position and the relative angle. Figure 1.5 shows the sensitivity of each measurement to the dip angle when the tool is moving across the boundary. The unit of y-axis is degree/degree and dB/degree for phase shift and attenuation curves respectively. We also conducted an experiment that shows the sensitivity of all four types of measurements to the position of the bed boundary. The unit of y-axis is degree/feet and dB/feet for the phase shift and attenuation respectively. The results are shown in Fig. 1.6.

As a result, based on the sensitivity studies, the combination of the measurements is capable to provide information on the dip angle and the layer boundaries, as well as the formation resistivity. This property is essential to the application of reservoir boundary mapping, formation evaluation, or well placement. However, the multi-layer earth structure is within the scope inevitably when implemented on the real drilling jobs [14]. Conventionally, the resistivity of surrounding earth formation can be derived from a lookup table that maps the measured curves to the resistivity profile. However, there is no direct interpretation of ultra-deep measurements since the model in interest is complex, which results in the complicated measurements. Hence, an advanced mathematical derivative is necessary [15], which helps infer a resistivity profile from the ultra-deep measurements.

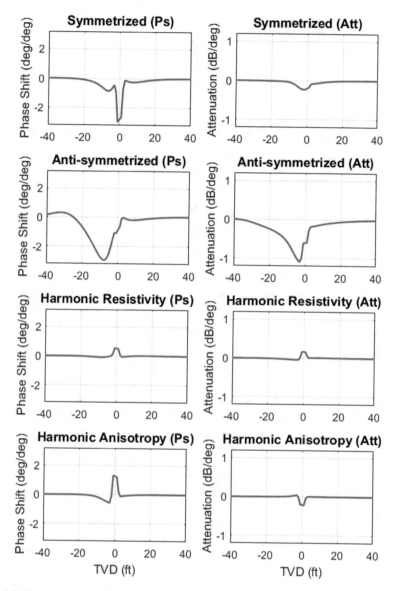

Fig. 1.5 The sensitivity of phase shift and attenuation of four types of curves to the dip angle when the tool moves across the interface

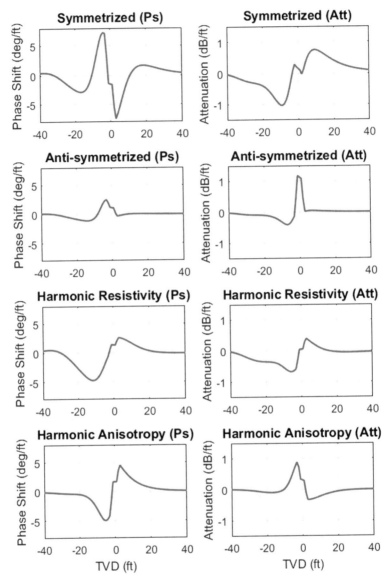

Fig. 1.6 The sensitivity of phase shift and attenuation of four types of curves to the layer boundary when the tool moves across the interface

1.3 Geological Inversion: Logging Data Interpretation

The interpretation of azimuthal resistivity measurements confronts a mathematical problem named inversion, which becomes the primary task in the full use of the advanced directional EM LWD service. A conventional inverse modeling is by applying deterministic optimization. A series of gradient-based algorithms are developed for fast inversion. In this section, an in-depth description of the inverse problem is introduced, followed by the introduction of existing solving methods. In the end, we will discuss the main challenge and introduce the main topic of this book.

1.3.1 Model Parameterizing and Inverse Problem

In the simulation of azimuthal resistivity tools, the forward and inverse modelings deal with the mappings between the earth model and the EM response. Modeling for the EM response is a full 3D problem taking account of the geological structure in the setting of directional EM LWD. However, there is an inevitable limitation in the computational resource that calculates the rigorous solution. Hence, a simplified scheme in a 1D model is more compatible with the context of real-time formation evaluation because of the available computational power and storage. It is assumed that the earth model is divided into multiple layers. To each layer, it has a constant physical property such as conductivity and anisotropy. Figure 1.7 demonstrates a typical 1D schematic that parameterizes the earth formations into multiple parallel layers.

A vector of geological parameters is used to describe an earth model in such a form

$$\mathbf{m} = [r_1, r_2, ..., r_k, z_1, z_2, ..., z_{k-1}]. \tag{1.2}$$

In the equation above, we are assuming a model vector $\mathbf{m} \in \mathbb{R}^{M \times 1}$. k denotes the number of layers in the 1D earth model. r_k is the resistivity value to the kth layer and z_{k-1} is the true vertical depth (TVD) of an interface between $k-1$ and k layers. We will keep these notations along the rest of this book. The EM responses are taken and measured along the drilling trajectory, and the 1D model at each measuring station can be inverted and grouped to draw a full 2D image of a cross-section resistivity profile.

There are two basic schemes to parameterize the model vector. We call the definition in Eq. 1.2 as model-based parameterization where two different types of parameters in physical domain and spatial domain are used to represent an earth model. There is another scheme named pixel-based parameterization, which discretizes the earth model equally into multiple thin parallel layers. Both schemes are shown in Fig. 1.8. The one can easily find that the model-based scheme reduces the number of overall unknown parameters greatly. Although there are some drawbacks and we

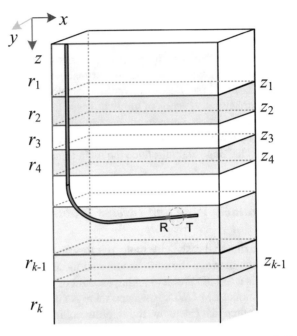

Fig. 1.7 A schematic diagram of a 1D earth model. The earth formation is divided into multiple parallel layers, each of which has a constant value of physical properties. The physical property is assumed to vary only in vertical depth

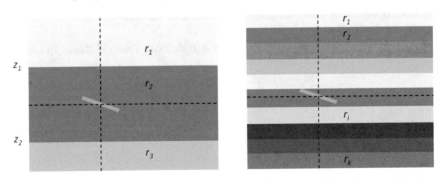

Fig. 1.8 Left panel: model-based parameterization where the earth model is defined by both physical and spatial values. Right panel: pixel-based parameterization where the earth model is equally divided into multiple thin layers

will discuss them in the next part, only model-based parameterization will be used throughout this book.

A standard definition for the inverse modeling is a mapping from observed data to the model parameters as $\mathbf{d} \to \mathbf{m}$. More specifically, it is the process to find the optimal model parameters $\mathbf{m}$, which can reproduce the observed data $\mathbf{d}$. We assume a collection of the measurements that $\mathbf{d} \in \mathbb{R}^{N \times 1}$ is a vector with a length of N. The objective is to minimize an error term between the forward response and the observation. The forward modeling denotes a computational model function $F :$ $\mathbb{R}^M \rightarrow \mathbb{R}^N$ which is designed to synthesize N responses from M model parameters. The forward transformation from the model parameters to the responses is calculated based on the 1D electromagnetic modeling, which is essentially solving Maxwell's equations with magnetic dipole sources in layered media [16, 17]. A good agreement between the response of the forward modeling and observed data can be reached if the inverted earth parameters come close to the actual scenario. The difference between the forward response and measurements is defined as data misfit $\mathbf{d} - F(\mathbf{m})$. Hence, a formal definition of the objective can be written as

$$\min_{\mathbf{m} \in \mathbb{R}^{M \times 1}} U = \|\mathbf{d} - F(\mathbf{m})\|^2 . \tag{1.3}$$

1.3.2 Deterministic Inversion

The Eq. 1.3 describes a nonlinear least-squares minimization problem. Many gradient-based optimization algorithms, such as the gradient descent method, Gauss-Newton method, and the Levenberg-Marquardt algorithm (LMA) [18, 19], have been well established to resolve such kind of problems. Although most of the optimization algorithms mentioned above are robust and have been implemented on many geological inverse problems [20], solving a non-convex optimization problem requires prior knowledge to make a feasible initial model as well as strong boundary conditions; otherwise, the solver can easily suffer from the local minima problem. A regularized objective function can be found in many applications of inversion [21]. Meanwhile, this type of inverse methods has been well studied in the earlier years and applied to the 1D layered geological inverse problems [22]. An Occam's inversion that firstly proposed by Constable et al. [23] can be one representative in the pixel-based inversion. It has a general objective function as

$$\min_{\mathbf{m} \in \mathbb{R}^{M \times 1}} U = \|\partial \mathbf{m}\|^2 + \mu^{-1} \|\mathbf{d} - F(\mathbf{m})\|^2 . \tag{1.4}$$

The regularized Occam's inversion seeks to minimize an objective that looks for the minimal data error, which measures the difference between the observed data $\mathbf{d}$ and forward estimation $F(\mathbf{m})$. Besides, it also demands a smoothness of the change of model parameters between layers $\|\partial \mathbf{m}\|^2$. The multiplier μ balances the data misfit and model smoothness. A significant reason for such a regularizing term is to concur the ill-posedness of a pixel-based inversion. However, the problem might become under-determined, where the number of solutions might be infinite. Therefore, the compensation to the interpretation of complex ultra-deep measurements is by posing additional prior knowledge or artificial assumptions.

The model-based inversion is the simplification when parameterizing an earth model. It parameterizes the earth model using a certain number of layers with associated resistivity value and the depth of each interface. It is effective when processing with simple scenarios to which only a few layers are involved. However, it can be easily under-parameterized or over-parameterized when handling complicated formation models.

The foundation of reservoir mapping/imaging is to infer the multi-layer structure as well as the complexity of the earth model, whereas the problem becomes contradictory if we have already possessed a strong understanding of the prior knowledge towards the object that we want to infer. People have been investigating the trade-off between artificial assumptions and the quality of inversion for a couple of years. A solution to reduce human intervention while maintaining the inverse quality is focused on optimizing the inversion workflow [24–27] instead of the principal methods. Once we saw the inadequacy of the deterministic approaches, two primary challenges are faced.

- The nonlinearity when solving the inverse mapping from data to model-based parameters.
- The complexity of a model, or the number of layers in use to describe the problem.

For the first point, the main difficulty in the model-based inversion comes from the high nonlinearity in the solution space. Secondly, avoiding under-parameterizing or over-parameterizing when handling model-based inversion becomes essential, which requires a self-parameterizing scheme for an optimal model definition. Beyond these two points, the challenging tasks raise the needs for a robust and efficient workflow of logging data processing and interpretation. Within the scope of challenges, this book aims at providing an effective solution to release the pains illustrated above, and presenting a systematic study of a class of methods. In next part, we briefly introduce an alternative approach to solve the inverse problems.

1.3.3 Statistical Inversion

In the setting of inversions, thinking through probabilistic formulation is one another approach. Instead of finding a unique solution of the inverted earth model, the objective becomes finding a probability distribution of the model space. The sampling method is developed to acquire an ensemble of predicted models from a probability distribution of the model parameters in a physical system [28]. This process is called a statistical inversion. Governed by Bayesian theorem, the posterior probability distribution (PPD) of the model parameters can be inferred through the likelihood of data and the prior probability of the model parameters. A class of sampling methods, Markov chain Monte Carlo methods, are used to sample this PPD. The samples drawn from PPD can reveal much statistical information. The statistics turns out to tell the probability of a potential model. The use of statistical inversion on the interpretation of geological/geophysical data is not new [10]. However, a comprehensive study has not been well accomplished, which results in an insufficient use of this method. On the other hand, the conventional statistical inversion has not stressed on the issues like efficiency, model selection, and parallelism.

In detail, the conventional statistical inversion makes the use of random-walk sampling, which results in a slow exploration rate. One effective solution is by introducing additional guidance for the sampler. The book investigates a hybrid scheme that combines gradient guidance with the process of random sampling [29–31]. We will also examine the feasibility and performance using this hybrid method to solve the inversion of ultra-deep measurements.

Secondly, to a model-based statistical inversion, the inference of model complexity, or the choice of the number of layers in an earth model, has been seldom stressed on [32]. Within the frame of Bayesian sampling, a generalized expression has been made, which turns the number of model layers to be a random variable. The sampling method, trans-dimensional Markov chain Monte Carlo (tMCMC), is introduced in detail. The book will illustrate how tMCMC infers the joint posterior distribution of model dimensionality and model parameters along the sampling process.

The statistical inversion offers a global searching capability, nevertheless, compromised significantly by the sampling efficiency, which is unacceptable to the real-time formation evaluation. The previous study [33] examined that tMCMC suffers even a lower acceptance rate than the fix-model MCMC inversion, which hinders the use of real-time formation evaluation such as geosteering and on-site reservoir mapping. It has to be admitted that such a probability solution makes efficiency worse [34]. Hence, the third study stresses on a high-performance sampling schematic [32, 35, 36]. A meta-technique, parallel tempering, is introduced beyond the Bayesian sampling frame and boosts the sampling performance dramatically.

1.4 Book Organization

This book focuses on the illustration of a class of statistical sampling methods and their applications on the inversion of geological logging data. The book is presented in five chapters. In the next chapter, the foundation of statistical inversion, its theory, and implementation, are illustrated. Chapter 3 discusses a particular scheme that combines the statistical inversion with local gradient assistant or known as the hybrid method. We will explain the theorem and methodology in detail. The fourth chapter posts an extension of conventional Bayesian inference that allows the interpretation of model complexity by using the trans-dimensional sampling method. In Chap. 5, we will introduce an inversion framework that combines Bayesian sampling with parallel tempering technique to enhance the sampling efficiency and take full advantage of the parallel computing system. The comprehensive study in this book aims to provide an efficient and complete statistical framework that searches for the optimal inverse model as well as the model complexity with minimized subjective assumptions. The book demonstrates multiple benchmark cases to examine the feasibility of each methods in the applications of formation evaluation.

References

1. Rodney PF, Wisler MM (1986) SPE Drill Eng 1(05):337
2. Clark B, Liiling MG, Jundt J, Ross M, Best D (1988) SPWLA 29th annual logging symposium. Society of Petrophysicists and Well-Log Analysts
3. Meyer WH (1995) SPWLA 36th annual logging symposium. Society of Petrophysicists and Well-Log Analysts
4. Bittar MS, Klein JD, Randy B, Hu G, Wu M, Pitcher JL, Golla C, Althoff GD, Sitka M, Minosyan V (2009) SPE Reser Eval Eng 12(02):270
5. Runge R (1974) Triple coil induction logging method for determining dip, anisotropy and true resistivity. U.S. Patent 3,808,520
6. Li Q, Omeragic D, Chou L, Yang L, Duong K (2005) SPWLA 46th annual logging symposium. (Society of Petrophysicists and Well-Log Analysts
7. Omeragic D, Dumont A, Esmersoy C, Habashy T, Li Q, Minerbo G, Rosthal R, Smits J, Tabanou J (2006) SEG technical program expanded abstracts 2006. Society of Exploration Geophysicists, pp 1630–1634
8. Edwards J (2000) SPWLA 41st annual logging symposium. Society of Petrophysicists and Well-Log Analysts
9. Beer R, Dias LCT, da Cunha AMV, Coutinho MR, Schmitt GH, Seydoux J, Morriss C, Legendre E, Yang J, Li Q (2010) SPWLA 51st annual logging symposium. Society of Petrophysicists and Well-Log Analysts
10. Seydoux J, Legendre J, Mirto J, Dupuis C, Denichou JM, Bennett N, Kutiev G, Kuchenbecker M, Morriss C, Yang L (2014) SPWLA 55th annual logging symposium. Society of Petrophysicists and Well-Log Analysts
11. Ezioba U, Denichou JM (2014) J Petrol Technol 66(08):32
12. Dong C, Dupuis C, Morriss C, Legendre E, Mirto E, Kutiev G, Denichou JM, Viandante M, Seydoux J, Bennett N (2015) Abu Dhabi international petroleum exhibition and conference. Society of Petroleum Engineers
13. Chen J, Wang J, Yu Y et al (2016) SPE J 21(04):1
14. Wang H, Shen Q, Chen J (2018) SPWLA 59th annual logging symposium. Society of Petrophysicists and Well-Log Analysts
15. Thiel M, Omeragic D (2018) SPWLA 59th annual logging symposium. Society of Petrophysicists and Well-Log Analysts
16. Huang M, Shen LC (1989) IEEE Trans Geosci Remote Sens 27(3):259
17. Zhong L, Li J, Bhardwaj A, Shen LC, Liu RC (2008) IEEE Trans Geosci Remote Sens 46(4):1148
18. Levenberg K (1944) Quart Appl Math 2(2):164
19. Marquardt DW (1963) J Soc Indus Appl Math 11(2):431
20. Huang S, Torres-Verdín C (2016) Geophysics 81(2):D111
21. Thiel M, Bower M, Omeragic D (2018) Petrophysics 59(02):218
22. Key K (2009) Geophysics 74(2):F9
23. Constable SC, Parker RL, Constable CG (1987) Geophysics 52(3):289
24. Dupuis C, Omeragic D, Chen YH, Habashy T (2013) SPE annual technical conference and exhibition. Soc Petrol Eng
25. Zhou J, Li J, Rabinovich M, D'Arcy B (2016) SPWLA 57th annual logging symposium. Society of Petrophysicists and Well-Log Analysts
26. Arata F, Gangemi G, Mele M, Tagliamonte R, Tarchiani C, Chinellato F, Denichou J, Maggs D, (2016) Abu Dhabi international petroleum exhibition and conference. Soc Petrol Eng
27. Campbell A, Beeley H, Bittar M, Walmsley A, Hou J, Wu HHM (2018) SPE annual technical conference and exhibition. Soc Petrol Eng
28. Mosegaard K, Tarantola A (1995) J Geophys Res Solid Earth 100(B7):12431
29. Shen Q, Wu X, Chen J, Han Z, Huang Y (2018) J Petrol Sci Eng 161:9
30. Gallagher K, Charvin K, Nielsen S, Sambridge M, Stephenson J (2009) Marine Petrol Geol 26(4):525

31. Martin J, Wilcox LC, Burstedde C, Ghattas O (2012) SIAM J Scient Comput 34(3):A1460
32. Shen Q, Chen J, Wu X, Han Z, Huang Y (2020) J Petrol Sci Eng 188:106961
33. Shen Q, Chen J, Wang H (2018) Petrophysics 59(06):786
34. Sambridge M (2013) Geophys J Int 196(1):357
35. Shen Q, Wu X, Chen J, Han Z (2017) Appl Comput Electromag Soc J 32:5
36. Lu H, Shen Q, Chen J, Wu X, Fu X (2019) J Petrol Sci Eng 174:189

Chapter 2
Bayesian Inversion and Sampling Methods

Abstract Statistic makes thinking a way through probabilistic formulation. As an alternative solution for many inverse problems, statistical inversion seeks for an ensemble of solutions instead of a unique one via a sampling process. The probabilistic equation is governed by the rule of Bayesian inference. In this chapter, we will introduce those fundamental concepts including Bayesian inference, Markov chain Monte Carlo method, as well as Metropolis-Hastings random-walk algorithm. Following these, we will show the magic that connects an inverse problem with Bayesian formula. One simple example will be presented to demonstrate the power of statistical inversion and the use on directional EM LWD data.

2.1 Introduction to Bayesian Inversion

In statistics, Bayesian theorem [1] describes the probability of an event given some prior knowledge related to this event and its condition. The formula is shown as

$$P(A|B) = \frac{P(B|A)P(A)}{P(B)}.$$

(2.1)

For example, event A is a person having lung cancer, while event B denotes a person who is a smoker. Hence, $P(A|B)$ targets to calculate the probability of a person having lung cancer given that this person is a smoker. On the right side of equation, the probability can be derived through the conditional probability $P(B|A)$ where the person is a smoker given he/she has lung cancer, and the prior information, the probability of having lung cancer, $P(A)$, and the probability of a smoker, $P(B)$. According to the Bayesian equation, it demonstrates a statistical update of a belief with coming evidence or more related information. The guidance of Bayesian theorem contributes to one prominent application of statistical interpretation or known as Bayesian inference [2].

Q. Shen et al., *Statistical Inversion of Electromagnetic Logging Data*,
SpringerBriefs in Petroleum Geoscience & Engineering,
https://doi.org/10.1007/978-3-030-57097-2_2

As one crucial technique, Bayesian inference stresses on the interpretation of probability for a hypothesis as the evidence becomes available. To make it clear, the same notation as defined in the first chapter is in use to formulate the equation below

$$p(\mathbf{m}|\mathbf{d}) = \frac{p(\mathbf{d}|\mathbf{m})p(\mathbf{m})}{p(\mathbf{d})}. \tag{2.2}$$

In Eq. 2.2, the hypothesis in research background is the earth model parameters $\mathbf{m}$ to be inferred, while the evidence comes from the measurements obtained as well as some prior knowledge that possessed before knowing the observed data. Recalling that observed data is $\mathbf{d} \in \mathbb{R}^{N \times 1}$, in the context of Bayesian inference, it derives the posterior probability of $\mathbf{m}$, given $\mathbf{d}$, through the prior probability and the likelihood of the observed data. $p(\mathbf{d}|\mathbf{m})$ is the likelihood function that reflects how likely the model $\mathbf{m}$ can reproduce the observed data $\mathbf{d}$. The likelihood function assesses the statistic of data error. It is built upon one assumption that the inverse mapping $\mathbf{m} = F^{-1}(\mathbf{d})$ may not include all factors that affect measurements. When considering noise in the observation, it is suggested that the noise is additive and comes from external sources, the relationship between observed outputs $\mathbf{d}$ and corresponding model parameters can be represented as

$$\mathbf{d} = F(\mathbf{m}) + \epsilon. \tag{2.3}$$

where ϵ denotes a random variable noise which follows an individual statistic. Experiments empirically suggest that additive noises usually follow a zero-mean Gaussian random distribution:

$$\epsilon \sim \mathcal{N}(0, \mathbf{C}_d). \tag{2.4}$$

Given model parameters $\mathbf{m}$ and observed data $\mathbf{d}$, the likelihood function can be deduced in a multivariate Gaussian distribution and takes a form like

$$p(\mathbf{d}|\mathbf{m}) = \frac{1}{\sqrt{((2\pi)^N |\mathbf{C}_d|)}} \times exp[-\frac{1}{2}(\mathbf{d} - F(\mathbf{m}))^T \mathbf{C}_d^{-1}(\mathbf{d} - F(\mathbf{m}))]. \tag{2.5}$$

In the equation above, $\mathbf{C}_d$ is the covariance matrix of the data error, $\mathbf{d} - F(\mathbf{m})$. It usually reflects the noise strength and is defined through the experiments of tool test. A numerical forward function $F(\cdot)$ calculates an estimation of 1D EM response given a certain earth model and tool configuration. It shares the exact form like data misfit term in Eq. 1.3, whereas in Eq. 2.5 it transforms the data misfit to the probability function of likelihood and describes the statistical property of data noise.

As a prior probability, $p(\mathbf{m})$ represents the belief of the target model before we take measurements and obtain the observed $\mathbf{d}$. The form can be various depending on the knowledge and information of a realistic task. However, in many cases without too much prior information, a wide uniform distribution is usually set for the probability density function (PDF) of $p(\mathbf{m})$ in a form that

$$p(m_i) = \begin{cases} \frac{1}{ub-lb} & lb \leq m_i \leq ub \\ 0 & \text{otherwise.} \end{cases} \tag{2.6}$$

The lower and upper boundaries, lb and ub, are usually set empirically based on a physical system. Meanwhile, The practice that uses a uniform distribution is to assume an unobstructed prior knowledge [3] and assure an unbiased boundary condition.

The last piece is the prior probability of data $\mathbf{d}$, which is also known as the marginal likelihood in a form that

$$p(\mathbf{d}) = \int_{\mathbf{m}} p(\mathbf{m}|\mathbf{d})p(\mathbf{m})d\mathbf{m}. \tag{2.7}$$

This probability remains constant for all possible models being considered. Therefore, the factor is the same for all hypotheses (the space of model parameters) and does not enter into determining the relative probability of the different hypothesis. This fact is essential when adopting sampling methods with a Metropolis-Hastings algorithm, which we will introduce in the next section. Consequently, Bayesian inference in Eq. 2.2 can be generalized in a relationship between posterior probability and likelihood function with the model prior as

$$p(\mathbf{m}|\mathbf{d}) \propto p(\mathbf{d}|\mathbf{m})p(\mathbf{m}). \tag{2.8}$$

The Bayesian inference offers a statistical approach searching for the target probability distribution. Instead of looking for a unique solution of model parameters, it returns only the probability. Sampling from the posterior distribution is one prominent way to understand the *shape*, or the statistical properties of target distribution [4], which provides an overall perspective in the solution space. It is called statistical inversion that sampling from the posterior distribution of model parameters given the observed data. The idea of statistic resolves the local minima problem in deterministic methods by searching for the statistical distribution of the earth model parameters. The approach of sampling can approximate the probability density function (PDF) of the target distribution, hence, derive the statistical properties such as mean value, the modes of distribution, or the variance.

Drawing samples from a posterior distribution $p(\mathbf{m}|\mathbf{d})$ is numerically incalculable due to the integration factor that involves a complicated forward function. In this setting, Markov chain Monte Carlo (MCMC) method appears to be a companion of Bayesian inference, which is a class of sampling algorithms that draws a series of samples from the posterior probability distribution (PPD) by constructing an invariant Markov chain whose stationary distribution is in the same with PPD. In the next section, we will have an in-depth explanation of the MCMC method as well as sampling using the Metropolis-Hastings algorithm.

2.2 Markov Chain Monte Carlo Method

MCMC is a strategy for generating a sequence of samples $\mathbf{m}^{(i)}$ from a certain distribution $p(\mathbf{m}|\mathbf{d})$ using Markov chain mechanism [5]. To reiterate the truth, direct sampling from an unknown distribution is infeasible, whereas evaluating the unnormalized probability value of $p(\mathbf{m}|\mathbf{d})$ is achievable. A Markov chain is a type of Markov process in a discrete state space. The current states of random variables depend solely on the previous states. Mathematically, the model state $\mathbf{m}^{(i)}$ in time (indexing) step i has a probability $p(\mathbf{m}^{(i)}|\mathbf{m}^{(i-1)}, ..., \mathbf{m}^{(1)})$ which is depended on the last state in time $i - 1$, to which it has

$$p(\mathbf{m}^{(i)}|\mathbf{m}^{(i-1)}, ..., \mathbf{m}^{(1)}) = T(\mathbf{m}^{(i)}|\mathbf{m}^{(i-1)}). \tag{2.9}$$

One should not confuse the conditional probability $p(\mathbf{m}^{(i)}|\mathbf{m}^{(i-1)}, ..., \mathbf{m}^{(1)})$ with the prior PDF defined in Eq. 2.6. The conditional probability here represents any probability distribution that we expect to sample. $T(\cdot|\cdot)$ is called stochastic transition kernel that $\sum_{\mathbf{m}^{(i)}} T(\mathbf{m}^{(i)}|\mathbf{m}^{(i-1)}) = 1$. It solely describes the transition, or movement, in a Markov process. Hence, constructing a Markov chain is a series process that creates an ensemble of random samples that in different states, and the process is controlled by the transition matrix $T()$. In MCMC, a fundamental requirement is the invariance of the distribution of the Markov chain. In other words, the chain will converge to stationary that the probability distribution over the state of $\mathbf{m}^{(i)}$ does not change along the stochastic process. To satisfy the stationary of the Markov chain, it requires two properties,

– *Irreducibility*: A Markov chain is irreducible if it can visit any state with non-negative probability in a finite number of steps.
– *Aperiodicity*: A Markov chain is aperiodic if it can return to any state at any time.

A sufficient condition that ensures the invariance of the Markov chain is it satisfies reversibility or detailed balance, where

$$p(\mathbf{m}^{(i)})T(\mathbf{m}^{(i-1)}|\mathbf{m}^{(i)}) = p(\mathbf{m}^{(i-1)})T(\mathbf{m}^{(i)}|\mathbf{m}^{(i-1)}). \tag{2.10}$$

Summing both sides over $p(\mathbf{m}^{(i-1)})$ and we can yield a form that

$$p(\mathbf{m}^{(i)}) = \sum_{\mathbf{m}^{(i-1)}} p(\mathbf{m}^{(i-1)})T(\mathbf{m}^{(i)}|\mathbf{m}^{(i-1)}). \tag{2.11}$$

Equation 2.11 provides an idea to design a sampler with a proper transition kernel which satisfies the detailed balance and can construct an invariant Markov chain. Consequently, the sampler will draw the samples from a target distribution. Meanwhile, it maintains the consistent distributions between the target and Markov chain.

2.3   Metropolis-Hastings Algorithm

Metropolis-Hastings Algorithm (MHA) is the most classical realization of MCMC drawing random samples sequentially from a probability distribution. Named after Nicholas Metropolis [6] who first proposed a sampling algorithm in 1953, Hastings [7] generalized later in 1970. To restate the advantage of MHA, it can draw samples from any probability distribution $p(\mathbf{m})$, provided a numerical function $f(\mathbf{m})$ that is proportional to the PDF function $p(\mathbf{m})$. This is crucial to apply Eq. 2.2 in the form of relationship denoted in Eq. 2.8, since the integration of model evidence is unable to calculate in practice. In general, MHA proposes a sequence of samples randomly. At each proposal step, a candidate model is generated randomly and accepted according to an acceptance probability. The workflow sampling from a posterior distribution $p(\mathbf{m}|\mathbf{d})$ is presented in Algorithm 1.

Algorithm 1 Metropolis-Hastings algorithm

Input: Initial model $(\mathbf{m}^0)$, chain length L

Output: samples $(\mathbf{m}^i)$ where $i < L$

  while $i < L$ **do**

   Perturb model parameters from a proposal function $q(\mathbf{m}^{i+1}|\mathbf{m}^i)$

   Obtain candidate as $\mathbf{m}^{i+1}$

   Calculate the acceptance probability

   $\mathcal{A}(\mathbf{m}^{i+1}|\mathbf{m}^i) = \frac{p(\mathbf{m}^{i+1}|\mathbf{d})q(\mathbf{m}^i|\mathbf{m}^{i+1})}{p(\mathbf{m}^i|\mathbf{d})q(\mathbf{m}^{i+1}|\mathbf{m}^i)}$

   Update the candidate

   if $\mathcal{A} < \mathcal{U}(0, 1)$ **then**

    $\mathbf{m}^{i+1} \leftarrow \mathbf{m}^i$

   end if

   Save $\mathbf{m}^{i+1}$ as one sample

  end while

The key to satisfy the detailed balance condition in Eq. 2.11 is the design of acceptance probability $\mathcal{A}(\mathbf{m}^{i+1}|\mathbf{m}^i)$ in Algorithm 1. Next, we will give a proof how MHA satisfy the detailed balance condition of MCMC.

Proof Define the target distribution to be sampled is $\pi(x)$, and the proposal function $q(x'|x)$. The acceptance probability is defined by $\mathcal{A}(x'|x) = min[1, \frac{\pi(x')q(x|x')}{\pi(x)q(x'|x)}]$

- If $\mathcal{A}(x'|x) < 1$, then $\mathcal{A}(x|x') = 1$, hence, $\mathcal{A}(x'|x) = \frac{\pi(x')q(x|x')}{\pi(x)q(x'|x)}\mathcal{A}(x|x')$
  which equals to $\pi(x)\mathcal{A}(x'|x)q(x'|x) = \pi(x')\mathcal{A}(x|x')q(x|x')$
- If $\mathcal{A}(x'|x) = 1$, then $\mathcal{A}(x|x') < 1$, hence, $\mathcal{A}(x|x') = \frac{\pi(x)q(x'|x)}{\pi(x')q(x|x')}\mathcal{A}(x'|x)$
  which equals to $\pi(x')\mathcal{A}(x|x')q(x|x') = \pi(x)\mathcal{A}(x'|x)q(x'|x)$

Assume $T(x'|x) = \mathcal{A}(x'|x)q(x'|x)$, for both conditions that $\mathcal{A}(x'|x) < 1$ or $\mathcal{A}(x'|x) = 1$, it satisfies the detailed balance condition in Eq. 2.11.

The choice of proposal function $q(x'|x)$ is arbitrary while the commonest proposal uses symmetrical Gaussian distribution as

$$q(x'|x) = \mathcal{N}(x, \sigma^2 I), \qquad (2.12)$$

where the covariance matrix Σ controls the jumping steps from the current state. The benefit of the symmetrical proposal is it simplifies the acceptance probability, which cancels the proposal probability term $q(\cdot|\cdot)$, because the reversible jump from state x' to state x has the same probability as the jump from state x to state x'. The acceptance probability reduces to the comparison between $\pi(x')$ and $\pi(x)$. Recall the posterior PDF in Eq. 2.2, the greatness of this step is that it avoids the calculation of evidence integration and merely calculates the likelihood and model prior functions. We call such a realization as random-walk MCMC using MHA. The name random-walk comes from the randomness of Gaussian proposal that perturbs the model variables to generate a new candidate. The tuning of MHA can influence sampling performance dramatically. One key hyper-parameter in MHA is σ, or the step size for a proposed state. We will examine the effect and tuning guidance of random-walk MCMC in the next section.

2.4 Examples of Using Random-Walk MCMC

2.4.1 Numerical Example

The example is started from a numerical test drawing samples from a banana-shape distribution. The unnormalized likelihood function has a numerical expression for the known target. Figure 2.1 demonstrates the contour of such a distribution.

Learned from Eq. 2.5, the constructed likelihood function consists of a forward evaluation, which is defined here as

$$\begin{cases} d_1 = & m_1 \\ d_2 = m_2 + (m_1^2 + 1). \end{cases} \qquad (2.13)$$

Now we assume the data $\mathbf{d}$ in Eq. 2.5 is observed as $\mathbf{d} = [0, 0]$. The target covariance $\mathbf{C}_d$ that controls the curvature of *banana* is set to $\begin{bmatrix} 1 & 0.9 \\ 0.9 & 1 \end{bmatrix}$. Hence, the global

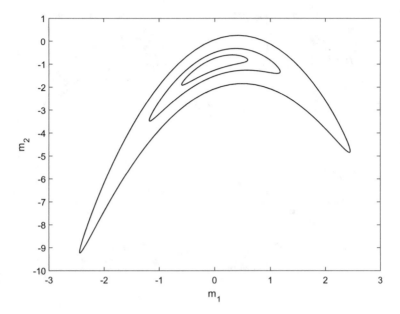

Fig. 2.1 Contour plot of a banana-shape distribution. There are two model parameters m_1 and m_2, where the global optimal is located at $\mathbf{m} = [0, -1]$

optimal can be found at $\mathbf{m} = [0, -1]$ according to Eq. 2.13. There is no prior setting of $p(\mathbf{m})$. In the following experiment, a sampler using random-walk MCMC is used to draw the samples directly from the defined likelihood. Recalling the algorithm 1, two user-defined hyper-parameters are needed. The total chain length defines the number of samples that the samper generates. In this example, 2000 samples are obtained. The second hyper-parameter is the proposal covariance (defined in Eq. 2.12) which controls the step size. It is essential to tune the performance of MCMC sampler. The following examples demonstrate the effect of running random-walk MCMC with different proposal covariance.

According to [8], the optimal acceptance rate for a random-walk MCMC is 0.234. It becomes one index to configure the step size and make the acceptance rate near the optimal. In the Fig. 2.2, the covariance matrix $\mathbf{C}_d$ is set to a diagonal matrix as $\begin{bmatrix} 1 & 0 \\ 0 & 1 \end{bmatrix}$. The performance shows a feasible tuning of the step size where the acceptance rate came to around 0.23. One can observe that the sample points cover this 2D region proportionally to the likelihood distribution. The left and bottom panels present the histogram of each model parameter. The envelope of histogram bins can be seen as the marginalized PDF of each model parameter.

The performance is compared with two different settings with larger or smaller proposal covariance. Figure 2.3 demonstrates the sampling effect, while the step size is too large or too small. In Fig. 2.3a, a small step size is chosen where $\sigma^2 = 0.1$. It shows a greatly increased acceptance rate while many effective samples are collected

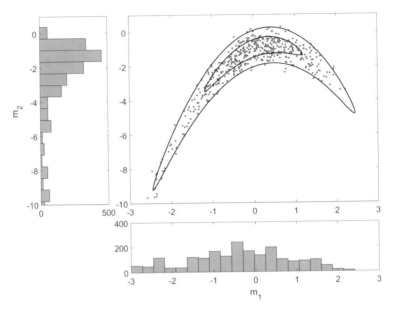

Fig. 2.2 Sampling a banana-shape distribution with an appropriate step size value, $\sigma^2 = 1$. The samples cover the distribution proportionally to the defined likelihood function. Left and bottom panels show the marginalized histogram of each model parameter

and presented in the high probability region. However, the movement of each iteration is too small, which impedes the sampler from drawing samples from the long tail regions. Therefore, the samples are insufficient to represent the entire distribution correctly. On the contrary, Fig. 2.3b shows a relatively large step size where $\sigma^2 = 10$. The acceptance probability is tiny, while the most proposed candidates are rejected. In this scenario, a longer chain length is necessarily required to obtain sufficient effective samples.

Figure 2.4 compares the chain behavior in three scenarios. The plot presents the change of the value of model parameter m_1 along chain iteration. It is usually called the trace plot. The upper plot shows a chain behavior in a small step size where the correlation between the neighboring samples is very high, which means the samples are gathering in a small region. The lower plot presents a chain with a large step size where many samples are rejected. This behavior causes insufficiency of useful samples and indicates that a longer chain length is necessary. Trace plot is a powerful tool to tune the performance of MCMC samplers, no matter what kind of sampling methods is used. In practice, a well-tuned MCMC sampler relies on the preliminary run to find out the optimal configuration. It differs from tasks. In the next section, we will implement the random-walk MCMC on the sampling of 1D earth model parameters and interpreting azimuthal resistivity measurements.

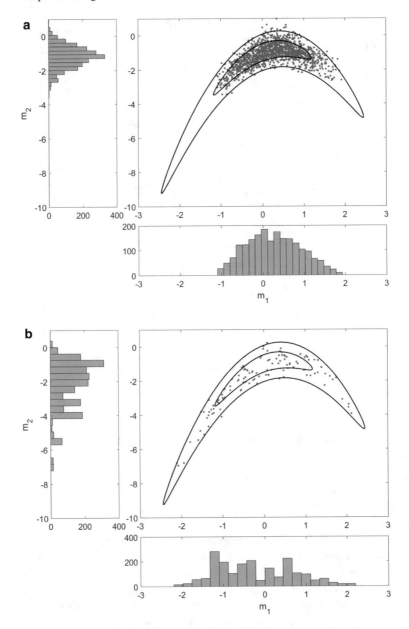

Fig. 2.3 Sampling a banana-shape distribution with small or large proposal covariance. **a** shows samples in a small step size. The sampler cannot jump out the high correlated region and sample the long tail of distribution. **b** shows samples in a large step size. Most of samples are rejected due to the low acceptance probability

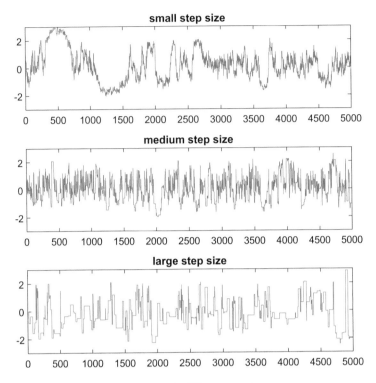

Fig. 2.4 Trace plots of model parameter m_1 in different step size

2.4.2 Synthetic Study of an Earth Model Inversion

Based-on 1D layered model presented in Fig. 1.7, a three-layer model is constructed with five model parameters $\mathbf{m} = [r_1, r_2, r_3, z_{up}, z_{down}]$ while the dip angle is fixed at a 90° which is generally applied to the horizontal scenarios. The tool is kept in the middle high-resistive layer to demonstrate a geosteering scenario. The target is to distinguish the upper and lower boundaries and maintain the drilling tool within the production zone and avoid breakthrough of boundaries. In this example, the resistivity value for three layers is 10, 50, and 1 ohm-m from up to down. We give some variance to the depth of layer interfaces $[z_{up}, z_{down}]$ to augment the number of logging station. Hence, a 2D earth section profile can be concatenated based on a series of 1D resistivity profiles at each logging station. Figure 2.5 shows the real model concatenated with 80 logging stations.

The central line indicates the tools navigation trajectory, which is fixed at a 90°. The inversion is conducted at each 1 ft in lateral depth (LD) while the test field is started from 0 to 80 ft in LD. The depth to either the upper or lower boundary is varying in different logging points. The largest depth to the boundary (D2B) is around 18 ft, and the smallest D2B is around 2 ft. The tool in use is configured

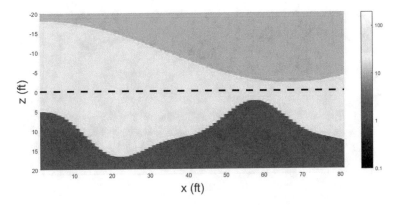

Fig. 2.5 A synthetic three-layer model with 90 logging points. The tool drills along the center dash line at a 90° dip angle

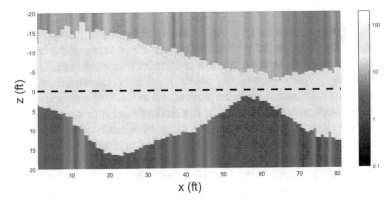

Fig. 2.6 The inversion result of three-layer model. 80 logging points are involved. To each point, the mean value of collected samples is used as the inverted earth model parameters

with multi-spacing and multi-frequency defined in Fig. 1.2. The longest T-R spacing spans around 7 ft while the frequencies cover 2 MHz, 400 kHz, and 100 kHz. The combination of different spacing and frequencies provides a multiple DoI as well as sensitivity. An artificial Gaussian noise is added on the measurement while 2σ value is 1.5 for phase shift and 0.25 for attenuation. At each logging point, 38 curves can be yielded where $\mathbf{d} \in \mathbb{R}^{38 \times 1}$. Given $\mathbf{d}$, the random-walk MCMC is used to sample from $p(\mathbf{m}|\mathbf{d})$. According to the preliminary run of Markov chain, the proposal covariance matrix is set to diagonal and keeps the acceptance rate at around 0.2. The total chain length is 5k, and the last one thousand samples are collected to infer the statistical properties of model parameters. Figure 2.6 demonstrates an inversion result.

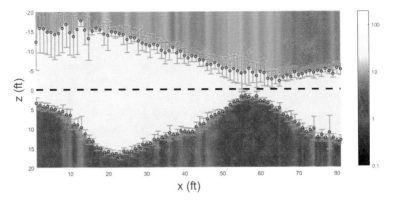

Fig. 2.7 The inverted earth model with boundaries uncertainty. The error bar denotes the variance range at each logging station

Compared with the true model in Figs. 2.5 and 2.6 shows a result that successfully reconstructed the earth model profile. We use the mean value of the collected 1k samples to represent the inverted value of earth model parameters **m**. The variance between logging points is introduced from the noise in the measurements. The error propagates from the data domain to the parameter domain. One significant advantage of statistical inversion is that the samples can reflect the error in the parameter space via statistics, which has an essential function for the uncertainty quantification. Figure 2.7 demonstrates the quantified uncertainty information of layer boundaries. In this case, we use the variance of collected samples as the quantified uncertainty value. The error bars overlapped on two boundaries show the uncertainty of parameters z_{up} and z_{down}. One can see that when the tool is far from the top boundary (the beginning part), the earth model cannot be well reconstructed. In other words, the uncertainty involved is high where the samples cover a large range of parameter region. The reason is that the sensitivity of the measurements is relatively poor when the tool is far from the layer interface. As the tool moves forward and the top boundary bends downward, the inversion can resolve both the upper and lower boundaries with decreased uncertainty.

From this example, we see the implementation of random-walk MCMC method on the statistical inversion of azimuthal resistivity measurements. It indicates an effective way to interpret the observed data and reconstruct the earth model resistivity profile. The Bayesian sampling approach conquers the local minima problem in conventional deterministic inverse methods through the probability perspective. However, drawing samples randomly is inefficient. The article tells that an average acceptance rate is around 0.2. In the next chapter, a hybrid method is introduced to overcome the issue of low efficiency in the random-walk MCMC method.

References

1. Bayes T (1991) MD Comput Comp Med Pract 8(3):157
2. Morris CN (1983) J Am Statist Assoc 78(381):47
3. Bodin T, Sambridge M, Tkalčić H, Arroucau P, Gallagher K, Rawlinson N (2012) J Geophys Res: Solid Earth 117:B2
4. Bishop CM (2006) Pattern recognition and machine learning. Springer
5. Andrieu C, De Freitas N, Doucet A, Jordan MI (2003) Mach Learn 50(1–2):5
6. Metropolis N, Rosenbluth AW, Rosenbluth MN, Teller AH, Teller E (1953) J Chem Phys 21(6):1087
7. Hastings WK (1970) Biometrika 57(1):97
8. Neal RM (2011) Handbook of Markov chain Monte Carlo 2(11):2

Chapter 3
Beyond the Random-Walk: A Hybrid Monte Carlo Sampling

Abstract Metropolis-Hastings algorithm realizes MCMC via random-walk sampling and recovers PPD through discrete samples. However, the slow convergence rate makes the sampling efficiency a big obstacle for any real-time data processing workflow. On the contrary, many deterministic optimizations follow a gradient update and have relatively fast searching speed compared with random move. One attractive realization is to combine two schemes, where people introduce the gradient as a moving force and combine it with a statistical sampling process. The fusion brings in the topic of this chapter, a hybrid scheme of MCMC, or named hybrid Monte Carlo sampling. In the following sections, we will introduce the concepts of hybrid Monte Carlo, Hamiltonian dynamics, and their mix, Hamiltonian Monte Carlo sampling method. We also use a group of examples to demonstrate the effect and advantage when applying HMC on the interpretation of directional EM LWD data.

3.1 Introduction to Hybrid Monte Carlo Method

The MCMC method guarantees an asymptotically exact solution for recovering the posterior distribution, though the computational cost is inevitably high and most MCMC algorithms suffer from a low acceptance rate and slow convergence with long burn-in periods [1]. Solving geological inverse problems by the statistical method is not new. The implementation of random-walk MCMC on the interpretation of sonic logging measurements has been realized [2]. However, the poor acceptance rate with long burn-in time is unbearable while applying this method to the interpretation of azimuthal resistivity measurements. In this section, the Hybrid Monte Carlo (HMC) is used to enhance the performance of naive random-walk sampling. HMC [3] united with Bayesian sampling and molecular dynamics provides a way that samples from the canonical density by simulating the dynamics of a physical system. Many experiments show that sampling performance is improved via the HMC method by avoiding random-walk behavior. In the following sections, we will first give a review of a specific physical system. It is followed by an introduction

© The Author(s), under exclusive license to Springer Nature Switzerland AG 2021
Q. Shen et al., *Statistical Inversion of Electromagnetic Logging Data*,
SpringerBriefs in Petroleum Geoscience & Engineering,
https://doi.org/10.1007/978-3-030-57097-2_3

of the HMC method that incorporates the Hamiltonian dynamics with MCMC sampling. We will verify the performance of HMC method with the synthetic model and demonstrate its advantages on solving statistical geological inverse problems.

3.2 Hamiltonian Dynamics

The HMC or so-called hybrid/Hamiltonian Monte Carlo replaces the proposal function of the random walk by the simulation of a dynamic process. The idea initially came from the observation of the motion of molecules and was well described by the Hamiltonian dynamics in the landmark paper by Duane et al. [3]. Generalized by Neal [4], HMC converges more quickly to the absolute probability distribution.

To understand the algorithm of HMC, first, we introduce the Hamiltonian dynamics with two sets of real variables $\mathbf{q}$ and $\mathbf{p}$. For a physical interpretation of these variables, the vector $\mathbf{q} = \{q_1, q_2, \ldots, q_n\}$ stands for the position of n molecules on a frictionless surface of varying height. The vector $\mathbf{p} = \{p_1, p_2, \ldots, p_n\}$ represents their current momentum. The corresponding energy functions, $U(\mathbf{q})$ and $K(\mathbf{p})$, also share the physical meaning as the potential energy and the kinetic energy, respectively. Suppose we are going to sample from the canonical distribution for the states of molecules, it becomes our interest to find the probability distribution of the molecules' position. In the Boltzmann probability, this canonical distribution $P(\mathbf{q})$ is related to the potential energy $U(\mathbf{q})$ in a form of

$$P(\mathbf{q}) = \frac{1}{Z_e} \exp(-U(\mathbf{q})), \tag{3.1}$$

where Z_e is a normalizing constant. Equation (3.1) bridges the view of energy and statistical probability given a state of the object. In the real world, vector $\mathbf{q}$ is a set of unknown parameters whose probability distribution is in interest. For each real variable q_i, there exists an associated variable p_i. Leaving out its physical meaning, the variable p_i is introduced independently to formulate a dynamical problem. The auxiliary random variable, $\mathbf{p}$ follow Gaussian distributions of the zero mean and unit variance. Therefore, the kinetic function is defined as

$$K(p) = \frac{1}{2} \sum_{i=1}^{n} p_i^2. \tag{3.2}$$

Hamiltonian function is concluded as an total energy function for variables $\mathbf{q}$ and $\mathbf{p}$, where

$$H(q, p) = U(q) + K(p). \tag{3.3}$$

It describes a dynamical system through the perspective of the energy, to which the partial derivatives determine how $\mathbf{q}$ and $\mathbf{p}$ change over continuous time t according to the Hamilton's equations:

$$\frac{dq_i}{dt} = \frac{\partial H}{\partial p_i}$$
$$\frac{dp_i}{dt} = -\frac{\partial H}{\partial q_i}.$$

(3.4)

The stochastic gradient equations above build up the foundation of HMC process. According to the definition of Hamiltonian function, some properties can be concluded and promise the feasibility when applying on the MCMC process. Starting from the first property, the Hamiltonian dynamics is reversible, which means from the current state at time t, $(\mathbf{q}(t), \mathbf{p}(t))$, to the state at time $t + s$, $(\mathbf{q}(t + s), \mathbf{p}(t + s))$, is objective and reversible. The reversibility of dynamic guarantees the desired distribution invariant during sampling. The second property is conservation, which denotes that the Hamiltonian function keeps invariant along time, which is

$$\frac{dH}{dt} = \sum_{i=1}^{n} \left[\frac{dq_i}{dt} \frac{\partial H}{\partial q_i} + \frac{dp_i}{dt} \frac{\partial H}{\partial p_i} \right] = 0.$$

(3.5)

The invariance of Hamiltonian reveals an equilibrium state for the Markov chain or a certain acceptance for Metropolis updates using a proposal found by Hamiltonian dynamics. It also ensures the construction of HMC sampling from an invariant distribution. Another property of the Hamiltonian function is that it preserves volume in the $(\mathbf{q}, \mathbf{p})$ space, known as the Liouville's theorem. One brief explanation is that if we apply the time mapping T to the points in some region R of the $(\mathbf{q}, \mathbf{p})$ space with volume V, the image of R under T will also have volume V [5]. This property guarantees the acceptance probability for the Metropolis updates without any influence by the change in its volume.

In order to sample the state of variables q_i and draw a distribution to reflect the statistical properties, the continuous-time Hamilton's equations must be approximated by discretizing the time with step ε. A commonly used scheme, the leapfrog method, is a way to simulate the state at a fixed time interval, dT, by alternatively moving $\mathbf{q}$ and $\mathbf{p}$ at halftime step ε, which minimizes the error introduced by the discretization. A single step to upgrade the position and momentum, $\mathbf{q}$ and $\mathbf{p}$ is shown as

$$p^{(k)}(t + \varepsilon/2) = p^{(k)}(t) - \frac{\varepsilon}{2} \cdot \frac{\partial}{\partial q^{(k)}} U(q^{(k)}(t))$$
$$q^{(k)}(t + \varepsilon) = q^{(k)}(t) + \varepsilon \cdot p^{(k)}(t + \frac{\varepsilon}{2})$$
$$p^{(k)}(t + \varepsilon) = p^{(k)}(t + \varepsilon/2) - \frac{\varepsilon}{2} \cdot \frac{\partial}{\partial q^{(k)}} U(q^{(k)}(t + \varepsilon)).$$

(3.6)

It starts with a half step update for the momentum variable $\mathbf{p}$, followed by a full step update for the target variable $\mathbf{q}$, and finally another half step for $\mathbf{p}$. With dT/ε times updating, the system will move to a new state. The discretized leapfrog process realizes a full update of a sample $\mathbf{q}$, governed by the Hamiltonian dynamics. The update of the auxiliary variable $\mathbf{p}$ is ignored since $\mathbf{p}$ are randomly drawn each time at the beginning of the leapfrog process.

3.3   Hamiltonian Monte Carlo Sampling

We now present a complete formula of Hybrid Monte Carlo Algorithm 2, which is used to sample from continuous distributions. There are two main processes at each iteration in the HMC algorithm. In the first part, new values of the momentum variables $\mathbf{p}$ are drawn randomly from their Gaussian distribution. Starting from the current state $(\mathbf{q}, \mathbf{p})$, an update is performed using the simulation of Hamiltonian dynamics and moves $(\mathbf{q}, \mathbf{p})$ in a distance by the leapfrog method to a proposed state $(\mathbf{q}^*, \mathbf{p}^*)$. In the second part, the Metropolis update is executed. The proposed state is accepted as the next state of the Markov chain at a probability, which is the same acceptance probability defined in Algorithm 1. If the proposed state is rejected, the current $\mathbf{q}$ will be kept for the next iteration.

Solving a geological inverse problem with HMC is similar to the random walk MCMC. Recall the statistical inversions introduced previously and we rewrite model parameters and observations by $\mathbf{m}$ and $\mathbf{d}$. Then the posterior distribution $p(\mathbf{m}|\mathbf{d})$ of the earth model parameters can be represented by the prior knowledge and likelihood function and is defined in Eq. 2.8. Now we replace all variable $\mathbf{q}$ by the familiar one, $\mathbf{m}$. According to Eq. 3.1, the corresponding potential energy $U(\mathbf{m})$ can be written as

$$U(\mathbf{m}) = -\log(Z_e \cdot p(\mathbf{m}|\mathbf{d})). \tag{3.7}$$

Following the Algorithm 2, a Markov chain is launched to draw samples of the model parameter $\mathbf{m}$ until the update reaches an equilibrium.

3.4   Using HMC for Statistical Inversion of Directional EM LWD Data

There are a few concerns about the implementation of HMC on the interpretation of azimuthal resistivity measurements. The primary one is the gradient of the potential function $U(\mathbf{m})$. The computational cost of the Jacobian matrix of forward function $F(\mathbf{m})$ is inevitably high compared to the random walk samplers. However, leveraged by the gradient-drifted property, where the proposal is always drifted to the region with a higher probability to be accepted, HMC has a much faster convergence rate

---

Algorithm 2 Hybrid Monte Carlo Method

Input: initial $m^{(0)}$, max chain length K

Output: $m^{(k)}$, where $k < K$

Initialize with arbitrary value $m^{(0)}$, step size ε, and leapfrog steps l

while $k \leq K$ **do**

  Randomly generate $p^{(k)}$ from $N(0, 1)$

  $m_0^{(k+1)} = m^{(k)}$

  $p_0^{(k+1)} = p^{(k)} - \frac{\varepsilon}{2} \cdot \frac{\partial}{\partial m^{(k)}} U(m^{(k)})$

  for $i = 1$ to l **do**

    $m_i^{(k+1)} = m_{i-1}^{(k+1)} + \varepsilon \cdot p_{i-1}^{(k+1)}$

    $p_i^{(k+1)} = p_{i-1}^{(k+1)} - \varepsilon \cdot \frac{\partial}{\partial m_i^{(k+1)}} U(m_i^{(k+1)})$

  end for

  $\mathcal{A}(m^{(k+1)}, m^{(k)}) = \min \left\{ 1, exp[\, U(m^{(k)}) - U(m^{(k+1)}) + K(p^{(k)}) - K(p^{(k+1)})]\, \right\}$

  Generate $\mathcal{A}_0$ from uniform distribution $\mathcal{U}(0, 1)$

  if $\mathcal{A}_0 < \mathcal{A}(m^{(k+1)}, m^{(k)})$ **then**

    keep $m^{(k+1)}$

  else

    $m^{(k+1)} = m^{(k)}$

  end if

  save $m^{(k+1)}$ in the chain

end while

---

than the random walk exploration. Also, the average acceptance rate of HMC is 0.65, while the random walk is around 0.2 [5]. The higher acceptance rate leads to a faster exploration of the target distribution and its statistical properties. In the meantime, the random mechanism inside the Hamiltonian dynamics differentiates HMC from deterministic inversions and avoids local optimums.

As a class of MCMC methods, choosing a suitable step size and leapfrog steps is crucial when implementing HMC. Together, the step size and leapfrog steps determine the length of the trajectory in fictitious time, εL. A large step size may result in unstable Hamiltonian dynamics due to errors introduced into the system by the leapfrog discretization. Too small a step size will produce little change of the state, waste computation time, and cause a slow exploration of the target distribution [5]. The selection of leapfrog steps is also a subtle work, and the distance to move within a simulation of the Hamiltonian may differ from one state to another state. In practice, similar to random-walk MCMC methods, a preliminary run of HMC is often required to determine a suitable choice of those hyper-parameters. An effective strategy is to allow a range of each hyper-parameter, and randomly choose a combination of those parameters for a single simulation of Hamiltonian dynamics. The randomness helps increase the overall possibility of fast exploration of the target distribution.

One another issue is that sampling in a single sequential chain tends to be unduly influenced by the slow-moving realization of iterative simulation, whereas multiple starting points can weaken the strong correlation [6]. Along with the increasing dimension of the sampling space, the multiple-chain sampling strategy has proven to help weaken the correlation of each sample and thus improving the possibility of convergence [7]. Multiple-chain HMC is implemented in this section to increase the overall sampling efficiency. This scheme is very suitable for parallel computing [8]. A simple strategy of a distributed MCMC method is built on the parallelization of multiple chains, which distributes the data and task of the HMC sampling to multiple processing units. There is a need to mention that this parallelism is lack of inter-sequences communication. We will discuss it with more details in Chap. 5. In this chapter, we make use of independent HMC samplers starting from different initial states. The experiments still show a positive impact of this scheme in exploring the target distribution.

In this section, we demonstrate the application of HMC method on solving the geological inverse problem and present the advantages of HMC compared to the traditional inversion methods. The examples are conducted based on the synthetic dataset generated by the simulated ultra-deep azimuthal resistivity LWD tool, which is configured with multiple working frequencies from the lowest 1 kHz to the highest 2 MHz. The corresponding transmitter-receiver spacing ranges from 800 in to 30 in. The assumption to DoI implies it can achieve up to 100 ft from the wellbore by using curves with a working frequency of 1 kHz. This simulated tool is presented and verified according to the most recent industrial development [9]. We will evaluate the proposed method on multi-layer earth models through several aspects, including the convergence plot, uncertainty analysis, and imaging of a multi-layer model.

3.4.1 Convergence and Acceptance Probability

The first example presents an examination of chain convergence. The metric to evaluate the performance of inversion is usually defined as the misfit between the actual value and the inverse value. There are two possible comparison ways, data misfit and model misfit. For a real-world task, the true solution of earth model parameters is never available while the comparison of data misfit is the only approach to evaluate the inversion performance. However, the non-convexity caused local minimum denotes a very small data misfit even the inverse model is tremendously different from a real one. Hence in our example, we use model misfit to investigate the performance of each method, since the real model parameters are synthetically constructed by ourselves.

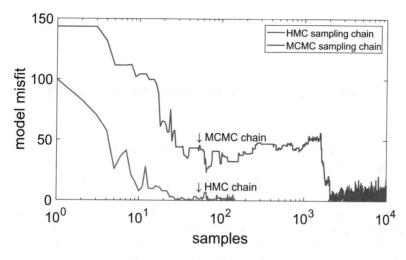

Fig. 3.1 Convergence of model misfit along the sampling steps.The upper MCMC chain converges after 2000 samples while the lower HMC chain converges faster within 30 samples with much higher acceptance rate

In this example, two instances are launched separately to compare the convergence speed of random-walk sampler and hybrid sampler. The goal is to sample a probability distribution of a three-layer model with five parameters. It consists of three resistivity values and two depth values of the layer interfaces. We use the model misfit, an L2 norm of the difference between the true solution and the inversion result, $\|\mathbf{m}_{true} - \mathbf{m}\|^2$, as an indicator to show the performance and chain convergence.

In Fig. 3.1, the result shows that the chain governed by HMC enters an equilibrium state after a burn-in period within 30 iterations, and the model misfit is near to zero, which means the samples are drawn from the target distribution successfully. However, the samples drawn by random walk MCMC are still outliers. The misfit of random-walk MCMC is always higher than the HMC samples along with the entire chain length. In other words, the random movement of parameters is with low efficiency to explore the parameter space given a limited chain length. Our experiment indicates that the model sampled by random-walk chain converges to true model after 2000 samples. This test agrees with that the statistical inversion by HMC method is more efficient than the random-walk MCMC method. HMC can draw samples precisely from the target distributions of model parameters. The other prominent improvement of HMC is that the acceptance rate is much higher, which guarantees the effectiveness in sampling the canonical distributions. MCMC suffers from the low acceptance rate shown in Fig. 3.1, where the blue line stays the same value for many iterations. This behavior indicates that the randomness is the only force to make the parameters change, whereas, to HMC method, a gradient-drifted manner helps to explore the target distribution faster. Hence, this explains that a much longer chain for random-walk MCMC is needed to obtain the same precision as the HMC achieves.

3.4.2 Synthetic Study of an Earth Model Inversion Using HMC

In the second example, a reconstruction of a synthetic three-layer model is conducted by applying HMC to 40 logging points in a 2000 ft horizontal well. The synthetic earth model, shown in Fig. 3.2a, is a three-layer model, where the resistivity values are 10 ohm-m, 50 ohm-m and 1 ohm-m from top to bottom, respectively. The central dash line indicates the tool's navigation trajectory. In this case, we assume the tool relative dip angle is fixed at 90°, and the drilling trajectory is horizontal. The depth to either the upper or lower boundary is varying on different positions, where the furthest boundary is up to 70 ft, and the closest one is 3 ft.

In Fig. 3.2b, we show the inversion result using HMC method with the uncertainty of interfaces similar to Fig. 2.7. The inverted model by HMC agrees with the true earth model with satisfactory performance. Beyond this, the error bar given by the

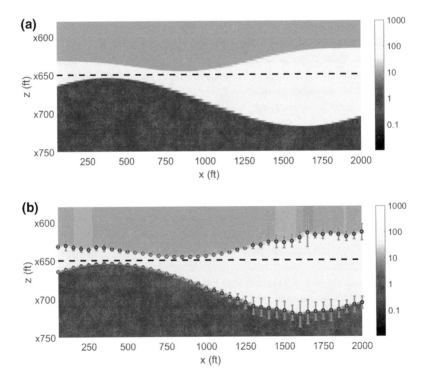

Fig. 3.2 The uncertainty evaluation of the inverse model

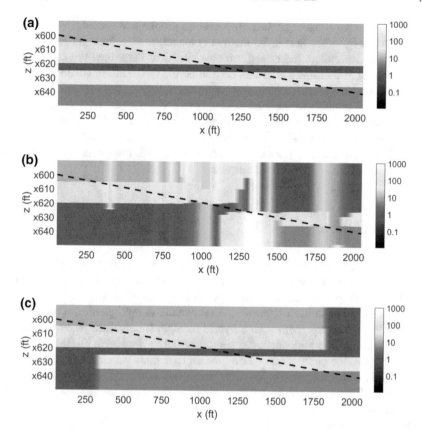

Fig. 3.3 A five-layer model. Comparison between deterministic method and statistical HMC method

HMC method on the two boundaries denotes the uncertainty of inversion results. In this case, the uncertainty becomes larger when the tool goes far from the boundary. The sensitivity of tool measurements becomes weak with more prominent data error, which propagates to the domain of model parameters. Overall, the statistical HMC inversion can reconstruct the earth model and returns the information about inverse uncertainty, which is meaningful to the real-time geosteering operation and helps keep the drilling tool within the desiring zone.

To the last example, we use a complicated five-layer model to verify the capability of HMC inversion, comparing with deterministic inversion. Figure 3.3a presents the real model, where the tool is drilling across from the top layer to the bottom along the dashed line. The formation resistivity changes from low to high and to low alternatively. Figure 3.3b presents inversion results by a deterministic method based on the Levenberg-Marquardt algorithm (LMA), a traditional gradient-based least-square optimization method. Generally, the deterministic inversions rely heavily on the initial guess. The outcome is unsatisfactory, and many inverse points suffer from incorrect recovery because of the non-convexity problem. We yield a hybrid result, as shown in Fig. 3.3c. The HMC method is launched on a multiple-point inverse model while we collect the drawn samples after 150 iterations. We calculate the mean value of 100 samples after this burn-in period as the intermediate results. We use LMA to refine the final inversion result and find the local optimal. From Fig. 3.3c, the HMC method presents the capability to find the region of global optimal of model parameters, and the refinement of the deterministic method help recover the model precisely. Although in some logging locations, we use a simplified three-layer model instead of a five-layer model to conduct the data interpretation, the overall performance of HMC inversion is auspicious.

3.5 Conclusion

In this section, a hybrid schematic is introduced that cooperates a gradient-driven Hamiltonian dynamics with MCMC sampling method. A significant improvement is observed compared with random-walk MCMC. The advantage can be summarized that the gradient information help improve the sampling efficient, thereby the acceptance probability. With such a hybrid method, it combines many advantages of both deterministic and stochastic approaches. The drawback of HMC can also be observed in the simulation of Hamiltonian dynamics which requires the calculation of the Jacobian matrix. To some cases, this computational expense cannot be ignored. Hence, the choice between methods should be considered in different working scenarios.

In the next chapter, we will stress on another issue that sampling more complicated earth models. It is crucial when interpreting ultra-deep measurements, where the layered structure within the scope is much more complicated. The choice of target model, or the number of layers, is essential. A brand new trans-dimensional schematic will be discussed.

References

1. Neiswanger W, Wang C, Xing E (2013) Asymptotically exact, embarrassingly parallel MCMC. arXiv:1311.4780 (2013)
2. Huang S, Yang Q, Matuszyk PJ, Torres-Verdín C (2013) SEG technical program expanded abstracts 2013. Soc Expl Geophysicists, 524–528
3. Duane S, Kennedy AD, Pendleton BJ, Roweth D (1987) Phys Lett B 195(2):216
4. Neal RM (1993) Probabilistic inference using Markov chain Monte Carlo methods. Department of Computer Science, University of Toronto Toronto, Ontario, Canada, Tech. rep
5. Neal RM (2011) Handbook of Markov chain Monte Carlo 2(11):2
6. Gelman A, Rubin DB (1992) Statist Sci 7(4):457
7. Shen Q, Wu X, Chen J, Han Z (2017) Appl Comput Electromag Soc J 32:5

8. Murray L (2010) Proceedings of neural information processing systems workshop on learning on cores, clusters and clouds, vol. 11
9. Bø Ø, Denichou JM, Ezioba U, Mirto E, Donley J, Telford J, Dupuis C, Pontarelli L, Skinner G, Viandante M (2015) Oilfield Rev 27(1):38

Chapter 4
Interpret Model Complexity: Trans-Dimensional MCMC Method

Abstract In the previous chapters, we witness the power of statistical inverse methods that used to sample from the posterior distribution of earth model parameters given the observed azimuthal resistivity measurements. The statistical inversion resolves the local minimum problem in the deterministic methods and tells the uncertainty of model parameters via the statistical distribution. However, the effect of using traditional MCMC methods is challenged when handling the ultra-deep azimuthal resistivity data. Besides, we have to answer the second question: how many parameters do we need to describe an earth model when solving model-based inversion? Stressing on the problems illustrated, a new method, trans-dimensional Markov chain Monte Carlo (tMCMC), is studied in this chapter. The method relaxes the problem dimensionality as an unknown parameter to be inferred. A novel algorithm named Birth-death algorithm will be introduced as one realization to implement tMCMC on an inverse problem. In this chapter, we will include one field case as example to demonstrate the miracle of tMCMC on the inference of model complexity.

4.1 Model Complexity and Self-parameterizing

To reiterate the issue of model complexity, we have learned that the earth models are usually built upon the fix-parameter assumption in the most methods that interpret LWD azimuthal resistivity measurements. Such a subjective choice of the model types commonly causes a systematic error and introduces parameter uncertainty since the inverted model can be either over-parameterized or under-parameterized. One primary solution is by using a pixel-based parameterizing scheme that discretizes the entire layered model into small grids which can be processed to avoid the presumption of model layers [1]. However, the pixel-based parameterizing makes the inverse problem under-determined while the number of unknown variables is more than the number of measurements. Although the constraint term can be applied to the objective like in Eq. 1.4 which uses a regularized Occam scheme, it becomes paradoxical while the assumption has been made that the layered model should be continuously changed along with the vertical depth. In practice, the Occam inversion should be penalized through prior model information when dealing with sudden changes in the earth

© The Author(s), under exclusive license to Springer Nature Switzerland AG 2021
Q. Shen et al., *Statistical Inversion of Electromagnetic Logging Data*,
SpringerBriefs in Petroleum Geoscience & Engineering,
https://doi.org/10.1007/978-3-030-57097-2_4

model. The foundation of reservoir mapping or imaging when using the ultra-deep services is to infer the multi-layer structure as well as the complexity of the earth model, whereas the problem becomes contradictory if we have already possessed a well understanding of the prior knowledge towards the object that we want to infer.

Stressing on the problems illustrated, a new method, trans-dimensional Markov chain Monte Carlo (tMCMC), is studied in this section. As a sampling method, tMCMC was firstly proposed by Green [2] as one generalization of traditional MCMC sampling methods. In this frame, the model dimension is an unknown random variable. The joint posterior distribution of model dimension and model parameters is inferred along the sampling process. The measurements will drive the sampling across different model dimensions. In other words, the probability of the models with different complexity and parameters are determined solely by the measurements. In a partitioned layered model, a simplified schematic named birth-death simulation has been used to simulate the tMCMC sampling process by inserting or removing a layer interface randomly [3]. Previous studies indicate that tMCMC can play an effective role that infers the model complexity when handling with geological inverse problems. The first geological application was presented by Malinverno [4] to invert DC (direct-current) resistivity sounding data. Subsequently, the applications were widely extended to a variety of geological inverse problems include the seismic tomography [5], geo-acoustic [6], or marine CSEM (controlled source electromagnetic) [7].

In the following sections, we will introduce the foundation of tMCMC which implement reversible jump MCMC method. One practical algorithm named birth-death method is illustrated in the next. We will examine the performance of tMCMC on some benchmark models.

4.2 Reversible Jump MCMC

In the 1D model-based inversion, a model is defined by several horizontally stacked layers (Fig. 1.7). A systematic error is always introduced in model-based inversion due to the deviation of the assumption of a model and the truth. Hence, it is important to interpret model complexity through the measurements instead of defining it by human-self. To learn the best representative of the earth model complexity, there should be a self-parameterizing scheme to determine the number of layers automatically. Within the frame of Bayesian inference defined in Eq. 2.8, Green [8] generalized it in a form that

$$p(\mathbf{m}, k|\mathbf{d}) \propto p(\mathbf{d}|\mathbf{m}, k) p(\mathbf{m}, k), \qquad (4.1)$$

where the number of layers k, as the representative of model complexity, is no longer given but a random variable and non-fixed. A trans-dimensional MCMC method is used to sample from the joint PPD of model parameters $\mathbf{m}$ as well as the number of layers k. A general way to create a stationary Markov chain is by using the MH algorithm, which is introduced as Algorithm 1. At each iteration, a candidate model is proposed from the current state $(\mathbf{m}, k)$. Unlike the fixed model proposal, tMCMC suggests a random move either within the current dimension or across different dimensions, which means k has a chance to change. In the setting of tMCMC, it has been shown [8] that the probability to accept the candidate state $(\mathbf{m}^*, k^*)$ should be

$$P(\mathbf{m}^*, k^* | \mathbf{m}, k) = min \left[1, \frac{p(\mathbf{m}^*, k^* | \mathbf{d}) q(\mathbf{m}, k | \mathbf{m}^*, k^*)}{p(\mathbf{m}, k | \mathbf{d}) q(\mathbf{m}^*, k^* | \mathbf{m}, k)} |\mathbf{J}| \right]. \qquad (4.2)$$

Similar to the general acceptance rule defined in Algorithm 1, a proposal function $q(a|b)$ presents the jumping probability from state b to state a. The major difference is the matrix J, which is the Jacobian of the transformation from the current state to candidate state. This abstractive term is introduced to account for the dimensional change during the sampling process that involves model transformation. Hence, the proposal process can be discussed regarding two general scenarios.

In a general state space, which involves no dimensional change, the random variable k does not change, and the proposal process reduces to a conventional random-walk proposal. Hence, a random multivariate Gaussian proposal function is used to generate a candidate in a form that

$$q(\mathbf{m}^*, k | \mathbf{m}, k) \sim \mathcal{N}(\mathbf{m}, \Sigma). \qquad (4.3)$$

The choice of Σ is the same as guided in the previous chapters where it needs preliminary run to determine a suitable step size for sampling. The jumping probability from $(\mathbf{m}, k)$ to $(\mathbf{m}^*, k)$ and its reversible jump has the sample probability, hence, $q(a|b)$ and $q(b|a)$ have been canceled. The determinant of Jacobian of the transformation is dismissed because no dimensional change happens. Therefore, the acceptance probability in Eq. 4.2 is reduced to

$$P(\mathbf{m}^*, k | \mathbf{m}, k) = min \left[1, \frac{p(\mathbf{m}^*, k | \mathbf{d})}{p(\mathbf{m}, k | \mathbf{d})} \right], \qquad (4.4)$$

which is the same as defined in random-walk MCMC.

The complicate things show up when dealing with trans-dimensional cases. Two terms in Eq. 4.1 are considered carefully. First, dimensional changes involved destroy the symmetric in the proposal. Therefore, the proposal probability cannot be canceled. The ratio between proposals should be determined, which requires us to calculate the probability in the reversible proposal. Second, the determinant of the matrix J is needed. A numerical answer to this abstract term is very difficult to learn. To most applications, the analytical solution can be derived. In the next section, a simulation named *birth-death* is introduced. It is a realization of tMCMC process which handles the model changes during sampling via an analytical process.

4.3 Birth-Death Algorithm

In the trans-dimensional cases, the birth-death simulation [3] is used to simplify the Jacobian matrix and make the sampling process analytically solvable. The simulation involves two moves; one is the *birth* move which increases the model dimension by one. Oppositely, the *death* move decreases the model dimension by one. Both moves have the same probability of happening. It has been shown that the Jacobian term is always unity in the birth-death simulation analytically [9]. Therefore, the calculation of MH acceptance probability is easier to be derived. In detail, the recipe of *birth* and *death* moves is as below. We also use Fig. 4.1 to graphically demonstrate this process.

birth move: The number of model layers increases by one as $k \rightarrow k + 1$. A randomly chosen depth is drawn uniformly from the investigation scope $[TVD_{min}, TVD_{max}]$. A new boundary is inserted at this location. Consequently, one layer is divided into two layers with the same resistivity value. Then, one of them is chosen randomly and its resistivity value is perturbed by a Gaussian proposal distribution.

death move: The number of model layers decreases by one as $k \rightarrow k - 1$. A randomly chosen boundary is removed from the current model. The associated neighboring layers are merged while the resistivity value is assigned as the mean value of the original two.

Calculating the proposal probability is a little fussy. For both birth and death steps, the change of model parameters can be made independently through the update of layer boundaries, $\mathbf{z}$, and corresponding resistivity value, $\mathbf{r}$ in the form that

$$\frac{q(\mathbf{m}|\mathbf{m}^*)}{q(\mathbf{m}^*|\mathbf{m})} = \frac{q(\mathbf{z}|\mathbf{m}^*)q(\mathbf{r}|\mathbf{m}^*)}{q(\mathbf{z}^*|\mathbf{m})q(\mathbf{r}^*|\mathbf{m})}. \tag{4.5}$$

For a birth step, the probability for each term in Eq. 4.5 can be defined. Specifically, giving a birth to a position has a probability

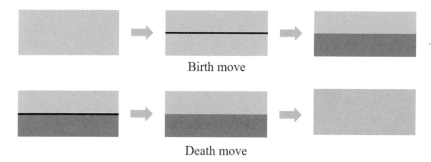

Birth move

Death move

Fig. 4.1 A graphical demonstration of birth and death moves on layered model

$$q(\mathbf{z}^*|\mathbf{m}) = 1/(N - k), \tag{4.6}$$

where we introduce an artificial variable N. It can be looked as the total number of grid within the scope. Each layer interface occupies a grid in this space. The probability to perturb the new-born resistivity is

$$q(\mathbf{r}^*|\mathbf{m}) = N(r_i, \sigma_i^2), \tag{4.7}$$

In its reversible jump, deleting an interface from $k + 1$ has a chance

$$q(\mathbf{z}|\mathbf{m}^*) = 1/(k + 1), \tag{4.8}$$

And removing the resistivity value when the cell is deleted has a chance

$$q(\mathbf{r}|\mathbf{m}^*) = 1, \tag{4.9}$$

because the layer does not exist once the boundary associated has been removed. Hence Eq. 4.5 can be determined as

$$\frac{q(\mathbf{m}|\mathbf{m}^*)}{q(\mathbf{m}^*|\mathbf{m})} = \frac{N - k}{(k + 1)N(r_i, \sigma_i^2)}. \tag{4.10}$$

Following the same way, the death move yields the proposal probability to be

$$\frac{q(\mathbf{m}|\mathbf{m}^*)}{q(\mathbf{m}^*|\mathbf{m})} = \frac{kN(r_i, \sigma_i^2)}{N - k + 1}. \tag{4.11}$$

It might be noticed that the artificial N has not been canceled. It is because we used an implicit expression of our prior distribution. Recalling Eq. 2.6, we use a uniformly distributed prior for all model prior. It does not matter when dealing with fixed model scenario since both proposal and its reversible terms can be canceled. Within the current context, we rewrite the prior probability explicitly as

$$p(\mathbf{m}) = p(\mathbf{r}, \mathbf{z}|k)p(k) = p(\mathbf{r}|k)p(\mathbf{z}|k)p(k). \tag{4.12}$$

Using the chain rule in conditional probability, the joint probability of $\mathbf{r}$, $\mathbf{z}$, and k has its form in Eq. 4.12 considering that the model variables $\mathbf{r}$ and $\mathbf{z}$ are independent. We can determine that within the preset upper and lower boundaries, $p(k)$ and $p(\mathbf{r}|k)$ are uniformly distributed and stay constant. For the prior of layer interfaces within an N grid system, it becomes

$$p(\mathbf{z}|k) = \left[\frac{N!}{k!(N - k)!}\right]^{-1}, \tag{4.13}$$

which represents that for k interfaces, there are $\frac{N!}{k!(N-k)!}$ possible configurations in N grid system. Hence we can rewrite the model prior in Eq. 2.6 by a new version that

$$p(m_i) = \begin{cases} \frac{k!(N-k)!}{N!\Delta} & within \\ 0 & otherwise \end{cases}, \tag{4.14}$$

where Δ represents the range between upper and lower boundaries. Adopting Eq. 4.14 for a birth step, the prior ratio can be expressed as

$$\frac{p(\mathbf{m}^*)}{p(\mathbf{m})} = \frac{k+1}{N-k}. \tag{4.15}$$

Hence, the term $N - 1$ Eqs. 4.10 and 4.15 will be canceled. Similarly, in the death step, the prior ratio is shown as

$$\frac{p(\mathbf{m})}{p(\mathbf{m}^*)} = \frac{N-k+1}{k}, \tag{4.16}$$

where the term $N - k + 1$ will be canceled either in Eqs. 4.11 or 4.16.

The process enables the transformation between neighboring dimensions and allows the sampling across different numbers of layers, hence, infers the model complexity. Algorithm 3 below presents the sampling process using tMCMC. By the end of the iterations, an ensemble of model samples in different dimensions is returned. One can find the most possible model and associated parameters according to the statistics.

4.4 Examples of Using Trans-Dimensional MCMC

In this section, tMCMC method is evaluated through several examples including synthetic single point evaluation, synthetic benchmark model inversion, as well as a field case study. The measurements provide multiple sensitivities towards formation resistivity, anisotropy, and earth interfaces with the maximal depth of investigation over one hundred feet. This section illustrates the implementation of birth-death schematic on the inversion of ultra-deep azimuthal resistivity data.

4.4.1 Inversion of Synthetic Multi-layer Models

The first example presents the inversion on a single-point model, where a set of synthetic observation $\mathbf{d}$ is generated from a five-layer model. Two hydrocarbon formations are prone to show beyond and below the current tool position. Figure 4.2 shows a schematic diagram. The tool relative position is at zero in TVD, and the boundary positions are set to -39, -15, 9, and 29 ft from up to down. Gaussian

---

Algorithm 3 Sampling PPD by trans-dimensional Markov chain Monte Carlo

---

Input: Initial model $(\mathbf{m}^0, k^0)$, chain length L, model change rate r

Output: chain samples $(\mathbf{m}^i, k^i)$ where $i < L$

  while $i < L$ **do**

    if $\mathcal{U}(0, 1) < r$ **then**

      if $\mathcal{U}(0, 1) < 0.5$ **then**

        Start birth move: $k \rightarrow k + 1$

        Generate z_{new} from $\mathcal{U}(TVD_{\min}, TVD_{\max})$

        Select r_{new} or r_{new+1} and perturb from $\mathcal{N}(r, \sigma^2)$

      else

        Start death move: $k \rightarrow k - 1$

        Randomly choose $z_{slc} \rightarrow [\,]$

        $r_{new} \leftarrow (r_{slc} + r_{slc+1})/2$

      end if

    else

      Perturb model parameters: $\mathbf{m}^{i+1} \sim \mathcal{N}(\mathbf{m}^i, \Sigma)$

    end if

    Obtain candidate as $(\mathbf{m}^{i+1}, k^{i+1})$

    $P = 1 \wedge \frac{p(\mathbf{m}^{i+1}, k^{i+1} | \mathbf{d}) q(\mathbf{m}^i, k^i | \mathbf{m}^{i+1}, k^{i+1})}{p(\mathbf{m}^i, k^i | \mathbf{d}) q(\mathbf{m}^{i+1}, k^{i+1} | \mathbf{m}^i, k^i)}$

    if $P < \mathcal{U}(0, 1)$ **then**

      $(\mathbf{m}^{i+1}, k^{i+1}) \leftarrow (\mathbf{m}^i, k^i)$

    end if

  end while

---

noises are randomly added to all measurements based on the general noise level, where the attenuation is with 0.05 dB, and the phase shift corresponds with 0.2° noise. The algorithm starts with a simple two-layer model, both initial resistivity and boundary position are randomly drawn from the prior setting. This manner gives the algorithm considerable flexibility and makes it independent of the initial model and the parameters. The total Markov chain length is set to 100k, where the first 10k samples are discarded as the burn-in period.

In this case, the forward responses consist of six frequencies from 1 kHz to 2MHz, where the low-frequency measurements (1 kHz, 4 kHz, and 16 kHz) provide

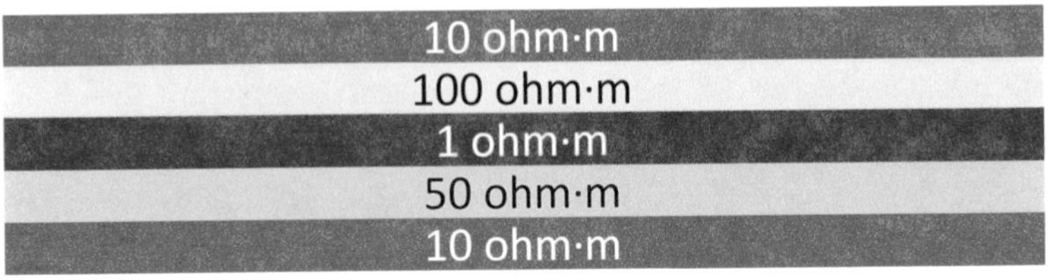

Fig. 4.2 A test five-layer model where the tool is in the center layer with 1 ohm-m resistivity

a capability of detecting far boundaries. High-frequency group (100 kHz, 400 kHz, and 2 MHz) contributes to resolving the local conductivity and near boundaries detection. All forward responses serve for the inverse process. Trans-dimensional inversions are carried out with some prior definition. Through many optimization approaches, such prior knowledge serves for constraint or boundaries conditions. Even so, this method adopts such convention, whereas with more flexible and loose conditions. The model allows dimensional change from two layers to twenty layers where $k_{min} = 1$ and $k_{max} = 20$ and the sampler has an equal chance to sample from all subspace of K. The boundary limit is set as the half of DoI, where $z_{min} = -50$ ft, $z_{max} = 50$ ft, and the logging tool is at $z = 0$. All born interfaces are drawn uniformly from (z_{min}, z_{max}). One additional constraint is the bed thickness. To avoid very thin bed, the minimal bed thickness is limited to 1 ft. Bounds for resistivity are empirically set to 0.1 ohm-m to 1000 ohm-m. To accommodate the physical interpretation, the resistivity is logarithmically drawn from a uniform distribution of -1 to 3.

We first present the behavior that model changes along with the tMCMC steps. The number of layers changed with steps, whereas the sampler kept sampling from the subspace of K mainly in the seven and eight layers model as Fig. 4.3 shows. Although the sampler started from a simple two-layer model, it jumped to sample from some more complicated structures after few iterations, which denotes that a simple earth model cannot support the true structure. Occasionally, the sampler jumped into higher dimensional regions with larger k. However, it returned to lower-dimensional models since the measurements drove the sampler to sample from the possible models who maximize a posterior probability of the geophysical structures. The histogram ranges from 2 to 20 as the prior of k regulated. The result shows that a seven or eight-layer model has the highest possibility to characterize the geological model. This conclusion is driven by the measurements data without any subjective assumption towards the earth model.

To interpret the statistical samples, a histogram is applied to all interface positions from all samples, as Fig. 4.4 shows. A considerable amount of information is revealed from this figure. First, a five-layer geo-structure with four potential layer interfaces can be easily identified via this histogram. With a longer bar, the interface has a higher chance to show at the corresponding depth. Second, uncertainty information can be extracted from this figure. A large spread of the bars denotes the significant uncertainty of the layer interface. The result agrees with this synthetic case where the resistivity logging tool is in the mid-layer at $z = 0$. The near boundaries close to the

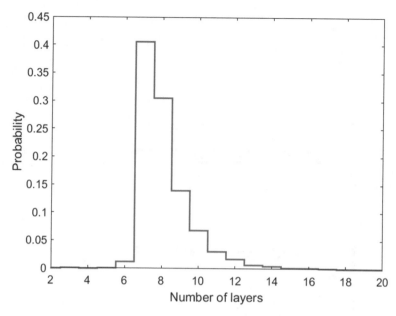

Fig. 4.3 The histogram of the number of layers from all samples. 7 to 8 layers have highest chance to characterize this model

logging tool are at -15 ft and 9 ft. Owing to the high resolution and short-spacing measurements, the near boundaries are detected with small uncertainties (spreading within a limited range). The further two boundaries are also detected since the deep-looking measurements provide such capability, however, with a larger uncertainty which results from multiple reflections of electromagnetic waves between boundaries and relatively lower sensitivities of deep-looking measurements.

Besides the information of interface positions, the resistivity and its uncertainty at each depth can also be extracted from all statistical samples. Figure 4.5 demonstrates one interpretation of layer resistivity. We firstly discretized resistivity in different models into (z_{min}, z_{max}) interval at 0.1 ft steps. Hence, for each row, or a certain depth, each sample answers a resistivity value. A history count is then conducted, where the color map is used to reflect the probability of different resistivity value. In Fig. 4.5, the warmer color shows a higher chance that the value of resistivity can be. Like the boundary histogram given in Fig. 4.4, a spread of resistivity value means an increased uncertainty. The result demonstrated is consistent with the model setup and the boundary histogram. Five different regions with corresponding resistivity ranges are clearly separated along with the depth. As described earlier, the tool is in the middle high-conductivity layer. Thanks to the conventional resistivity measurements, which can tell the formation resistivity around the tool, the uncertainty of its resistivity is near to zero, where the figure shows the color concentrates in a single line with maximal probability. The prediction of resistivity in other remote layers gains uncertainties due to the measurements noise and lower sensitivity. The figure

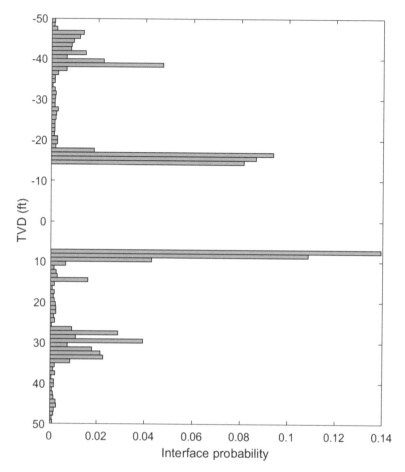

Fig. 4.4 The statistics of model interface. The longer bar represents a higher possibility a layer boundary has at a certain depth

also tells that the resistivity value near the boundary has a large uncertainty, where the value spans in multiple regions.

Following by this single point evaluation, we now use a benchmark model to demonstrate a full model inversion, which consists of 221 logging points. A Standardization of Deep Azimuthal Resistivity (SDAR) benchmark model is in use, which is created by the SPWLA Resistivity Special Interest Group (Rt-SIG) [10]. B1 model is selected for this example, which is a five-layer model shown in Fig. 4.6. It demonstrates a reservoir-mapping and imaging task where the center dash line denotes the drilling trajectory. The LWD tool is drilling downwards at a fixed 82° dip angle. The target is to recover the geological structure as precise as possible using ultra-deep azimuthal resistivity measurements. The major challenges come from the resolution

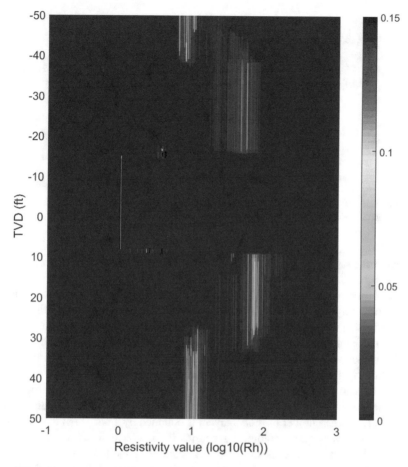

Fig. 4.5 The statistics of model resistivity. The color denotes the probability of the resistivity value in logarithm. The spread of the bright region represents a possible resistivity value with an uncertainty

of the first approaching thin bed (6 ft) as well as recovering the earth details within a large scope (±100 ft) from the borehole.

The synthetic ultra-deep tool is configured with a system of one-transmitter and two-receiver BHA subs that provide two spacing of 30 ft and 75 ft approximately. Three working frequencies are available to both T-R pairs, which covers 2 kHz, 6 kHz, and 12 kHz. Given a combination of one spacing and frequency in a single channel, four different types of measurements are constructed based on the formation coupling tensors [11, 12]. The combination of four types of measurements provides sensitivity to different geological features, including the formation boundary, relative dip angle, bulk resistivity, and anisotropy. Each type of measurements returns the values of attenuation and phase shift. Hence, in this example, 48 measurements are

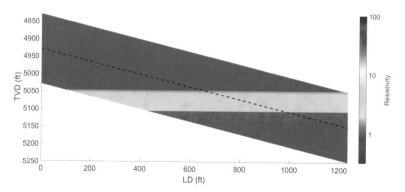

Fig. 4.6 SDAR B1 model with 221 logging points. The LWD tool drills along with the center dash line and across multiple geological structures

available at each logging station. All measurements are added with Gaussian noise that $2\sigma = 1.5°$ for phase shift and $2\sigma = 0.25$ for attenuation.

The setting of boundary conditions is one of our research interests. It serves as the prior probability in the context of statistical inversion. One purpose of this study is to investigate a new self-parameterizing scheme. Hence, there is no presumption of a particular geological structure or reference model. In our experiments, only upper and lower bounds are applied to each type of parameters. Uniform distributions of the earth model parameters are defined based on universal physical knowledge. In details, the resistivity value r_i is bounded by 0.1 ohm-m to 1000 ohm-m in logarithm scale; the interface depth is bounded within the tool DoI and varies depending on the job scenarios. The number of model layers is bounded by 2 and 20 in this example. This setting covers most geological scenarios in HAHZ applications. There is no other prior assumption besides the relaxed boundary condition mentioned before. The initial state is set to a homogeneous conductive model. The resistivity is 1 $\Omega \cdot m$. The sampler will interpret the measurements and infer the statistics of model parameters as well as the model complexity. To each logging point, the maximum chain length is set to 25k where the last 5k samples are collected to evaluate the sample statistics.

Figure 4.7 demonstrates the inversion result using the mean of the last 5k samples for each logging point. Due to the noises in the measurements, the reconstructed section profile is noisy, whereas the major geological structures can be distinguished. To the initial stage from 0 to 200 ft in LD, tMCMC sampler suggests a simple conductive formation where the measurements have little sensitivity to the thin layer. It can be recovered and distinguished while the tool is in 400 to 800 ft in LD. The main resistive hydrocarbon is detected and reconstructed after drilling across the thin bed and below the tool position. tMCMC sampler returns a multi-layer structure when drilling between two resistive layers. This example reflects the capability of tMCMC method in inferring model complexity driven by the measured data.

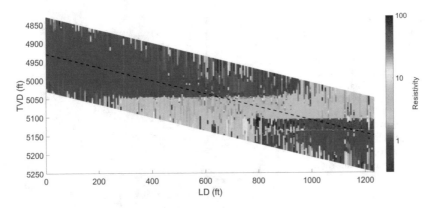

Fig. 4.7 The comparison between the local formation resistivity profile (RT) and the mean value of the statistical inversion result when the tool locates at 3022 m in TVD. A sharp interface at the depth of 3038 m is precisely recovered

4.4.2 A Field Case

A field example from a subsea gas well of Wheatstone liquefied-natural-gas project in Western Australia is used to assess the proposed measurement interpretation schematic [13]. The ultra-deep azimuthal propagation resistivity service was adopted for the geo-stopping task. The main objective is to halt the drilling immediately above a hydrocarbon-bearing zone to top-set casing at that point without penetrating into the productive core of a gas interval. The well must also intersect the horizontal top of the target zone at the inclination of 45–60°. The conventional LWD systems with shallow DoI are insufficient to provide "looking ahead" ability as required in this project. The azimuthal propagation resistivity service became the positive candidate for the task.

A two-receiver system was used in this geo-stopping operation. The distance between transmitter-sub and receiver-sub are 10.38 m and 24.24 m, respectively. The operating frequency is ranged between 2 kHz and 24 kHz. Figure 4.8 presents the basic logs in WST-1D well of the paper published by Upchurch et al. in 2016 [13]. The measured depth, the true vertical depth (TVD) and the local resistivity from the conventional LWD propagation tool are shown in the 1st, 2nd, 3rd and 4th track respectively. The wellbore inclination is 56°. As we can see, a sharp boundary presents at a depth of 3038 m in TVD. The formation resistivity of the overburden layer and the reservoir layer are about 4 ohm-m and 100 ohm-m respectively. The high sharpness on the interface as well as the resistivity contrast poses the ideal condition for the distance-to-boundary (D2B) inversion of the azimuthal resistivity measurements.

In this study, we conduct tMCMC processing on the azimuthal resistivity measurements at the depth 3022 m in TVD. The tool locates in conductive formation with 16 m above the reservoir interface, and it was halted in the real job. The total

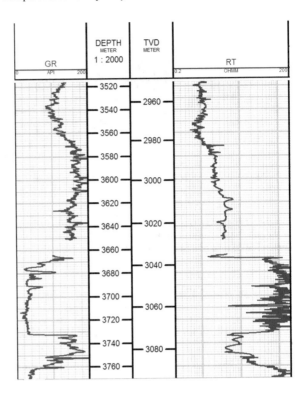

Fig. 4.8 The basic logs in WST-1D well. GR, measured depth, TVD and local resistivity are shown in the 1st, 2nd, 3rd and 4th track respectively. A sharp boundary presents at the depth of 3038 m in TVD. The formation resistivity of the overburden layer and the reservoir layer are about 4 ohm-m and 100 ohm-m respectively

number of the measurements used for the statistical interpretation is 38. In this exercise, the minimal bed thickness is set to 5 ft, and the number of bed is set between 2 to 10. Again, the last 50k samples from tMCMC sampling are selected to evaluate the statistical properties of the inverted parameters, such as the formation boundary position and the resistivity profile.

The probability maps of the layer interface and the distribution of the resistivity profile are plotted in Fig. 4.9. In Fig. 4.10, the comparison between the local formation resistivity profile (RT) and the mean value of the statistical inversion result from tMCMC are overlapped on the probability graph. Again, a multi-layer structure is suggested from tMCMC processing. A sharp interface at a depth of 3038 m is precisely recovered, which suggests the deep-looking measurements are sensitive to the reservoir top. However, the significant difference between the RT and the mean value of the statistical inversion result raises cautions on using the inverted resistivity for water saturation mapping.

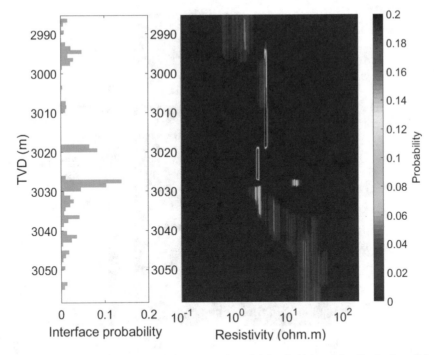

Fig. 4.9 The probability maps of the layer interface (left-hand side) and the distribution of the resistivity profile (right-hand side) when the tool locates at 3022 m in TVD

4.5 Conclusion

In this chapter, a generalized Bayesian formula is introduced that allows the number of model parameters to be a free random variable. The experiments present the feasibility of trans-dimensional sampling approach that makes the inference of model complexity possible. This is essential when interpreting the ultra-deep measurements where the geological features in reservoir-scale are more complicated than in traditional applications. However, the trade-off is obvious. The previous sections stress the problem of sampling efficiency, thereby considered a hybrid schematic. The situation becomes worse when it moves to a trans-dimensional case. It is observed that the average acceptance rate decreased to around 0.1 when dealing with model changes. A proposed model candidate in different dimensions may cause a higher rejecting rate due to the low likelihood. It is against the baseline of real-time interpretation because a single long chain may require over 100 k iteration to obtain sufficient samples which can reflect a true model distribution. In the last example, the noisy reconstruction of the model section profile also results from insufficient sampling. Taking the number of samples into consideration, a better choice is to make the sampling process parallel using the strategy of multiple samplers which can be deployed on multiple computational units hence reduce the single-chain length. In the next

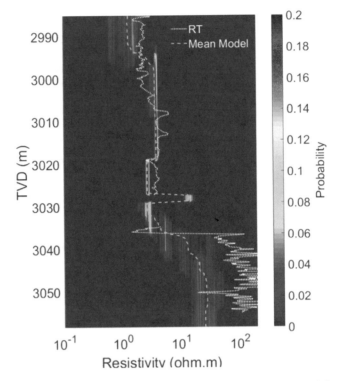

Fig. 4.10 The comparison between the local formation resistivity profile (RT) and the mean value of the statistical inversion result when the tool locates at 3022 m in TVD. A sharp interface at the depth of 3038 m is precisely recovered

chapter, an important technique is introduced that allows multiple Markov chains to work together and increase the sampling efficiency.

References

1. Thiel M, Bower M, Omeragic D (2018) Petrophysics 59(02):218
2. Green PJ (1995) Biometrika 82(4):711
3. Stephens M (2000) Ann Statist 28(1):40
4. Malinverno A (2002) Geophys J Int 151(3):675
5. Bodin T, Sambridge M (2009) Geophys J Int 178(3):1411
6. Dettmer J, Dosso SE, Holland CW (2010) J Acoust Soc Amer 128(6):3393
7. Ray A, Key K (2012) Geophy J Int 191(3):1135
8. Green PJ (2003) Oxford statistical science series, 179–198
9. Denison DG, Holmes CC, Mallick BK, Smith AF (2002) Bayesian methods for nonlinear classification and regression. Wiley
10. Wilson G, Xie H, Yu Y (2019) Udar and dar benchmarks models—forward and inverse modeling study. https://www.spwla.org/SPWLA/Chapters_SIGs/SIGs/Resistivity_/Resistivity.aspx (2017). Online; Accessed March 14, 2019

11. Li Q, Omeragic D, Chou L, Yang L, Duong K (2005) SPWLA 46th annual logging symposium. Society of Petrophysicists and Well-Log Analysts
12. Dong C, Dupuis C, Morriss C, Legendre E, Mirto E, Kutiev G, Denichou JM, Viandante M, Seydoux J, Bennett N (2015) Abu Dhabi international petroleum exhibition and conference. Society of Petroleum Engineers
13. Upchurch ER, Viandante MG, Saleem S, Russell K (2016) SPE Drill Compl 31(04):295

Chapter 5
Accelerated Bayesian Inversion Using Parallel Tempering

Abstract Applying tMCMC resolves two major problems we post at the beginning. One is the local minima problem and the other is a model selection problem. However, the observation tells us an inadequate performance when sampling a complex model. The decreased sampling efficiency is due to the dimensional changes. Hence, one possible solution comes to make MCMC methods more scalable and to be deployed on a high-performance computing system. The idea brings people to look for a parallel version of MCMC algorithms. In this chapter, we will introduce such a parallel schematic, which can be combined with any type of MCMC sampling method. We will first introduce the fundamental concept of tempering. And then we use the combination of tMCMC with parallel tempering to form up a complete workflow for solving a fast statistical inverse problem.

5.1 Introduction to Multiple Chains

There was a debate on the choice of one long run in MCMC or many short runs. It had been discovered that running multiple MCMC sampler independently from different starting states helped diagnose long burn-in period [1]. Both random-walk and hybrid sampling indeed face non-convergence in poor tuning, where the sampler keeps drawing samples from a mode in a region of distribution. It is fatal to the high-dimensional geological inverse problem since the mean of samples usually plays the role of an inversion result. Although the stochastic process guarantees a chance to get rid of a high correlated region, it might need extremely long iterations to jump out; hence, the entire burn-in period could be insufferably long. Launching multiple Markov chains starting from different regions of target distribution may help cut down a highly correlated chain into individual shorter chains and boost the exploration of the target distribution. Enlightened by this idea, the posterior distribution can be sampled through the joint probability distribution of every Markov chains launched. This scheme has been used in our hybrid Monte Carlo sampling and presented in the last example shown in Fig. 3.3c.

© The Author(s), under exclusive license to Springer Nature Switzerland AG 2021
Q. Shen et al., *Statistical Inversion of Electromagnetic Logging Data*,
SpringerBriefs in Petroleum Geoscience & Engineering,
https://doi.org/10.1007/978-3-030-57097-2_5

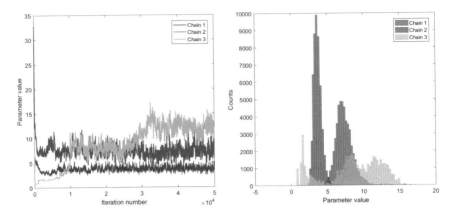

Fig. 5.1 Running three independent Markov chains and sample from a posterior distribution. The left panel shows the trace plot of one model parameter while the right panel shows the histograms for three groups of samples

Running multiple samplers simultaneously with different, usually randomized, starting distribution makes the application of parallel computing available such as Map-Reduce scheme [2], and in high-performance computing (HPC) scheme [3]. It seems to be harmless for the sampling efficiency since all Markov chains are deployed and iterated on their computational units. An example can be shown by drawing samples from a posterior distribution and visualizing the trace plot of one specific model parameter.

Figure 5.1 demonstrates the scenario we described. A joint probability distribution can be recovered through the samples from three chains instead of a single one. There are multiple modes which lead the second chain to sample a distinct region of its target distribution, whereas the joint distribution will finally converge to a true target. However, the goal may not be guaranteed. Interestingly, another result suggests a different interpretation of this idea.

From Fig. 5.2, one can hardly tell if the chains are sampling from the target (converged) or not. There are three individual modes in their joint distribution shown by the histogram, which suggests this model parameter has a very high chance to be either mode value. The trace plot demonstrates a poor chain mix. One way that obtains the inversion result is to break down the joint distribution and select one from three modes by evaluating data misfit. However, this process is against the original purpose using a statistical sampling method. Running multiple deterministic optimizers by different initial guesses could be cheaper and more efficient regarding statistical sampling approach.

No doubt run multiple chains at the same time can boost the performance of exploring the joint distribution. However, it plays less meaning running multiple independent chains simultaneously [4]. To figure out a better solution, one may take a look at both Figs. 5.1 and 5.2. It would be more effective if chains can exchange information about the regions they sample. According to [5], inference from a pos-

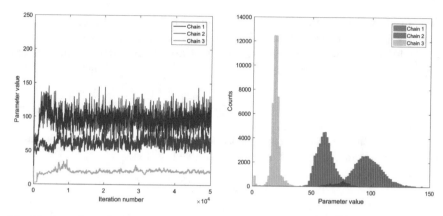

Fig. 5.2 An embarrassing case that three chains reflect three individual modes

terior distribution using multiple chains relies on effective communication during iterations. The nested Markov chain sequences are the key to a fast exploration of the target joint distribution. Multiple identical chains starting from different region helps less avoid high correlated regions. To assist the main objective, some of the samplers could be auxiliary, and the common goal is to assist the main sampler to explore a multi-modal distribution successfully. In this chapter, we will discuss a mega-technique, tempering, which is combined with multiple chains sampling methods and designed for this purpose.

5.2 From the View of Energy: Tempering Methods

Tempering technique was initially a stochastic process that evaluates the physical status of a dynamical system based on the energy variable. It has a history in global optimization and sampling. A well-known global optimization method, simulated annealing [6], is one successful algorithm to simulate the energy annealing process in the physical system. It was inspired by annealing in metallurgy, which simulates a cooling process and follows the change in its physical system. Another tempering technique, parallel tempering, was initially proposed by Geyer [7]. Multiple replicas of Markov chains run in different 'temperature' and exchange information with each other. The chains in higher temperature acquire a better dynamic, which assists the whole sampling system to explore the target distribution faster and more efficiently.

In this section, we will show how to connect a general objective function to its energy form. In the context of optimization, if one is seeking for a global optimum of an objective $\phi(x)$, it is equivalent to instead sample its PDF in a form that

$$\pi(x) = \frac{1}{C} e^{-\phi(x)}, \tag{5.1}$$

where C is a constant to normalize PDF. Equation 5.1 shares the same form and meaning as Eq. 3.3 defined in Chap. 3. It considers the energy within a system where the global minimum of the objective $\phi(x)$ will maximize the probability defined Eq. 5.1. The tempering works on its energy form by introducing a temperature factor, T like

$$\pi(x) = \frac{1}{C} e^{-\frac{\phi(x)}{T}}. \tag{5.2}$$

We use a straightforward example to demonstrate the effect of introducing a temperature factor. Assuming there is an objective function,

$$\phi(x) = (x^2 - 1)^2, \tag{5.3}$$

its corresponding unnormalized PDF, according to Eq. 5.1 has an expression that

$$\pi(x) = e^{-(x^2-1)^2}. \tag{5.4}$$

In Eq. 5.4, two modes appear at $x = -1$ and $x = 1$ respectively. Sampling $\pi(x)$ is not easy without a well-tuned configuration. Because the sampler tends to be sampling in the region of one peak around $x = 1$ or $x = -1$ with limited chance to jump out the high correlated region. If the proposal variance is not large enough, the proposed candidate will be regularly rejected according to the acceptance probability. Applying the form described in Eq. 5.2, tempering is a transformation of the original distribution by adding a power term T as

$$\pi(x, T) = e^{\frac{-(x^2-1)^2}{T}}. \tag{5.5}$$

The tempering effect can be seen through the shape of PDF in different temperature. Figure 5.3 shows a class of unnormalized PDF of $\pi(x, T)$ where $T = 2^i$ and $i = 0, 2, 4, 8, 10$. When $T = 1$, the $\pi(x, 1)$ is the original distribution in interest. As the temperature increases, the bimodal distribution becomes relatively flattened with two maxima less pronounced, although the shape of $\pi(x, 1)$ has been preserved. The limit of $\pi(x, T)$ as $T \to \infty$ becomes a uniform PDF where all modes are flattened. The phenomenon indicates that the locations of all modes are not changed with increasing temperature while the difference in parameter domain becomes less important. In this bimodal example, it is easier to sample a flattened distribution than the original one. Although releasing samplers to sample tempered versions is much easier to make a full exploration within parameter domain, the samples gathered in higher temperature where $T \neq 1$ can no longer reflect the true distribution in which we are interested.

The question becomes to design a strategy which brings the high dynamicity of samplers in high temperature to the sampler working on the original task. Simulated annealing is one solution, which allows a single sampler to draw samples from a tempered distribution where the initial T value is very high. It follows the temperature ladder and cools down along the iterations. Finally, the sampler will converge to a

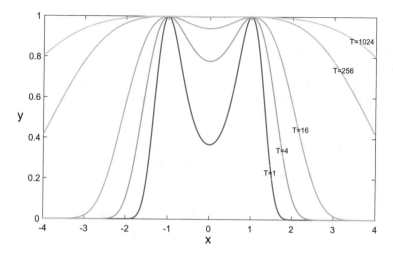

Fig. 5.3 A bimodal distribution and its tempered forms. The red curve shows a bimodal distribution with two modes at $x = -1$ and $x = 1$. The gray curves show its tempered forms in different T values

global optimum. However, this scheme cannot conquer the multi-modal problem without an appropriate cooling strategy. A single chain iterates to only one mode after too fast cooling down. Hence, neither can the entire distribution be reflected. Most importantly, annealing cannot allow multiple chains with a communication strategy. In the next section, a parallel tempering scheme is introduced, which will stress on this issue.

5.3 Parallel Tempering

Parallel tempering, or replica MCMC, adopts the idea of tempering to increase the sampling dynamic, which leads to a faster exploration of the target distribution. Unlike simulated annealing, multiple samplers are running in parallel. A temperature ladder is created where $T \geq 1$. In the context of Bayesian inference, a posterior probability distribution (PPD) is written in $p(\mathbf{m}|\mathbf{d})$. A *cold* chain is sampling the target PPD while the other hot chains are sampling its tempered forms, $p(\mathbf{m}|\mathbf{d})^{T^{-1}}$. The ensemble of chains together provides a way to sample the joint distributions $p(\mathbf{m}|\mathbf{d}, T_i)$ where $i = 1 : n$. We define this joint distribution as

$$\pi(x) = \prod_{i=1}^{N} c(T_i)^{-1} p(\mathbf{m}_i|\mathbf{d})^{1/T_i}, \tag{5.6}$$

where $c(T_i) = \int p(\mathbf{m}_i|\mathbf{d})^{1/T_i} d\mathbf{m}$ are the marginalized term as a constant for a given temperature. The significance of parallel tempering is that two states are selected randomly from two chains in different temperature and swapped as

$$(\mathbf{m}_i, T_i), (\mathbf{m}_j, T_j) \rightarrow (\mathbf{m}_j, T_i), (\mathbf{m}_i, T_j). \tag{5.7}$$

The swap in two states creates communication between chains. It gives the *cold* chain an opportunity to obtain a dynamic, which allows it to move to another region explored by hot chains, where a swap happens between $\mathbf{m}_0$ and one of the other states. However, the detailed balance of two swapping chains breaks since the transition kernel of the involved Markov chains cannot satisfy the reversibility condition. In order to maintain the equilibrium of the *cold* chain that keeps the stationary distribution as desired, the M-H acceptance rule is applied to determine if this swap is accepted or not. Assuming that two chains i and j are chosen randomly and is accepted by the probability

$$P(i, j) = min\left[1, \left[\frac{p(\mathbf{m}_j|\mathbf{d})}{p(\mathbf{m}_i|\mathbf{d})} \right]^{T_i^{-1}} \times \left[\frac{p(\mathbf{m}_i|\mathbf{d})}{p(\mathbf{m}_j|\mathbf{d})} \right]^{T_j^{-1}} \right]. \tag{5.8}$$

The Eq. 5.8 is derived from the original MH acceptance rule that

$$\mathcal{A}(x'|x) = min[1, \frac{\pi(x')q(x|x')}{\pi(x)q(x'|x)}], \tag{5.9}$$

where the swapping proposal is symmetrical, hence the jumping $q(|)$ terms are canceled and

$$\begin{aligned} \frac{\pi(x')}{\pi(x)} &= \frac{c(T_i)^{-1}p(\mathbf{m}_j|\mathbf{d})^{1/T_i}c(T_j)^{-1}p(\mathbf{m}_i|\mathbf{d})^{1/T_j}}{c(T_i)^{-1}p(\mathbf{m}_i|\mathbf{d})^{1/T_i}c(T_j)^{-1}p(\mathbf{m}_j|\mathbf{d})^{1/T_j}} \\ &= \left[\frac{p(\mathbf{m}_j|\mathbf{d})}{p(\mathbf{m}_i|\mathbf{d})} \right]^{T_i^{-1}} \times \left[\frac{p(\mathbf{m}_i|\mathbf{d})}{p(\mathbf{m}_j|\mathbf{d})} \right]^{T_j^{-1}}. \end{aligned} \tag{5.10}$$

Once the swap is accepted, the states that contain the information of the current regions are exchanged, which brings a chance for both chains in different temperature to explore a broader region of distribution. Swapping states is significant to the cold chain to sample the target distribution entirely instead of a portion of one high correlated region since the hot chains can provide extra dynamics to the entire system. The swapping can happen at any iteration along the sampling process. There is a trade-off between a maximal use of parallel computing system and the necessary communication between chains. A common implementation is by letting each chain run independently and exchange information every certain number of iterations. Algorithm 4 shows the algorithm of parallel tempering. It should be noted that as a meta-technique, parallel tempering can be applied to any MCMC method. Hence, in the next chapter, multiple examples are demonstrated, and we will use a frame that combines parallel tempering and tMCMC to demonstrate the performance.

Algorithm 4 Parallel tempering

Input: Initial model $\mathbf{m}^0$, chain length L, number of chains N, swap interval S

Output: chain samples $\mathbf{m}^i$ where $i < L$

  while $i < L$ **do**

    for i to $i + S$ **do**

      Proceed all chains in parallel

    end for

    randomly select (a, b) from $1 : N$

$$P = 1 \wedge \left[\frac{p(\mathbf{m}_b|\mathbf{d})}{p(\mathbf{m}_a|\mathbf{d})}\right]^{T_a^{-1}} \times \left[\frac{p(\mathbf{m}_a|\mathbf{d})}{p(\mathbf{m}_b|\mathbf{d})}\right]^{T_b^{-1}}$$

    if $P > \mathcal{U}(0, 1)$ **then**

      Swap sample states: $\mathbf{m}_a \leftrightarrow \mathbf{m}_b$

    end if

  end while

5.4 Examples of Using Parallel Tempering

5.4.1 Numerical Example

We start the first example from sampling a bimodal distribution. The unnormalized PDF of this distribution is defined in Eq. 5.4. The shape and form of this bimodal distribution can be found in Fig. 5.3. In this test, five Markov chains are launched to sample the original distribution as well as its tempered forms which follows a temperature ladder $T = [32, 8, 2, 1]$. In this example, we compare two cases where the first one is without chains communication. In the second case, the cold chain is coupled with chains in tempered distributions and exchange state with each other randomly during each iteration. Figure 5.4 demonstrates the first case.

Without communication between the cold and hot chains, the cold one in violet only sampled one region around $x = -1$ while two hot chains ($T = 32, 8$) were able to move around the two modes. The histograms of four uncoupled chains present the shapes of distributions sampled by four chains. Although the tempered distributions can be well recognized with two major peaks, the original distribution sampled by the cold chain is wrongly concentrated on one peak around $x = -1$. In the next step, we introduce communication by exchanging chain states randomly. The performance can be seen in Fig. 5.5.

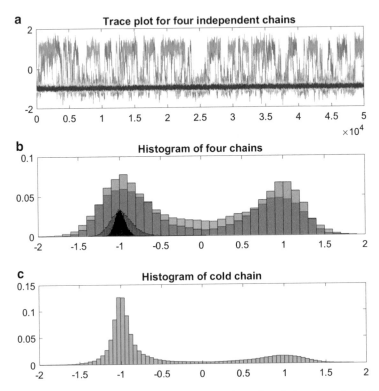

Fig. 5.4 From up to bottom, **a** The trace plot for four independent chains where the violet curve denotes the cold chain. **b** A comparison of sampled distributions by four independent chains. **c** The wrongly sampled target distribution, where most samples gather around -1

The sampling performance has been greatly improved when allowing communication between chains. In a chance, the cold chain exchanges the current state with high dynamical chains and obtains the opportunity to move around the entire distribution. It gets an equal chance to sample from two peaks at $x = -1$ and $x = 1$. This numerical example demonstrates the advantage of parallel tempering, which breaks the high correlation within a single chain by introducing the temperature dynamics. In the next example, we present the power of parallel tempering in the interpretation of azimuthal resistivity measurements and see how it can save a bad-tuned chain from wrong sampling and enhance the overall imaging quality.

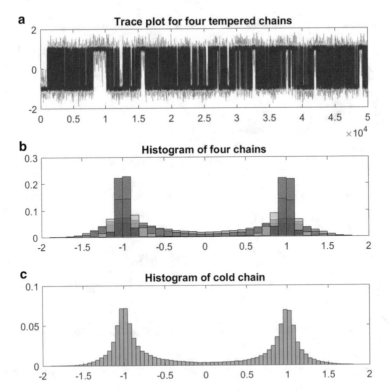

Fig. 5.5 From top to bottom, **a** The trace plot for four independent chains where the violet curve denotes the cold chain. **b** A comparison of sampled distributions by four independent chains. **c** The truly sampled target distribution, where most samples gather around two peaks at −1 and 1

5.4.2 Interpretation of Ultra-Deep Data Using Parallel Tempering

5.4.2.1 A Single Point Test: Random-Walk MCMC with Parallel Tempering

We test parallel tempering from a simple case first, where a single three-layer model is evaluated using random-walk MCMC. It has five model parameters where $r = [2, 50, 5]$ and $z = [-20, 40]$. The tool is at the origin where TVD $= 0$ and is with the same configuration as defined in the previous chapter. The lowest 2 kHz working frequency gives tool ultra-deep capability. We use the same starting state at a homogeneous conductive formation and iterate four random-walk samplers in parallel. Similar to the previous example, we compare two scenarios running multiple chains independently and with parallel tempering. A slight difference is that four chains are identically sampling from the same distribution and with no tempering scheme to the first case.

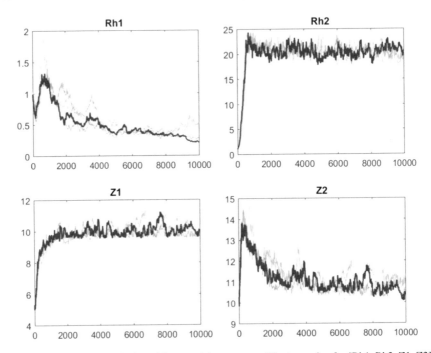

Fig. 5.6 The trace plots of selected four model parameters. The true value for [Rh1, Rh2, Z1, Z2] is [5] where all chains are moving into a wrong region. The violet curves denote the first chain we select

Figure 5.6 shows the trace plots of selected four model parameters. The true value for [Rh1, Rh2, Z1, Z2] is [5] (we use the absolute value of Z1 to represent the distance to the upper interface), whereas all chains are moving into the wrong region. Four chains behave similarly with the same step size and starting point. The example proves our point that independent chains running in parallel have no use for the improvement of sampling performance. A better solution for multiple independent chains is using a randomized starting state, which may help them better explore of other regions in the posterior distribution of model parameters.

This situation can be greatly changed when applying parallel tempering on random-walk samplers. The trace plots are shown in Fig. 5.7. The gray curves in the background denote chains who are sampling tempered posterior distribution. In about 6k iterations, a successful swap between the cold chain and high dynamic chain happened, and the state of cold chain transformed into another region of this distribution. This communication helped sampler move closer to the true earth model value. It is important for an inverse problem where sampling from a local mode is not our target.

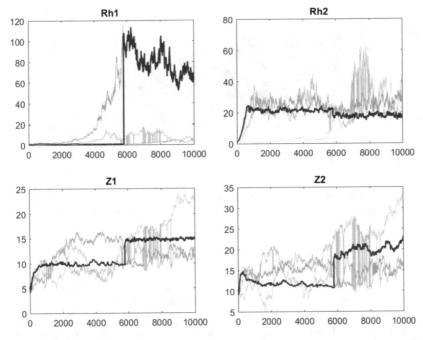

Fig. 5.7 The trace plots of selected four model parameters. Gray curves denote the chains sampling in tempered PPD. A successful swap happened on cold chain at around 6k iterations, which moved sampler out of local high correlated region

5.4.2.2 Benchmark Model Test: SDAR A3 and B1

In the last example, we evaluate the performance using parallel tempering to boost trans-dimensional MCMC method. As the two most powerful techniques, the combination of tMCMC and parallel tempering enables an efficient and robust interpretation of ultra-deep measurements as well as the model complexity. To illustrate the performance of this framework, in this section, we demonstrate several examples of benchmark models. We evaluate the overall performance of the earth model inversion through imaging quality, processing efficiency, and the robustness to relaxed boundary conditions.

Recall the tool configuration explained in Chap. 4, the synthetic ultra-deep tool is configured with a system of one-transmitter and two-receiver BHA that provides two spacings of 30 ft and 75 ft approximately. Multiple working frequencies are available to both T-R pairs, which covers 2 kHz, 6 kHz, and 12 kHz. 48 measurements are outputted at each logging station. The synthetic Gaussian noises are added to the observed data. The noise level is set to $0.5°$ as one σ to the phase shift and 0.1 dB as one σ to the attenuation. Boundary conditions are set to [0.1, 1000] ohm-m for resistivity value, and ±100 ft from tool depth for the locations of layer interfaces. There is no other prior assumption besides the relaxed boundary condition mentioned

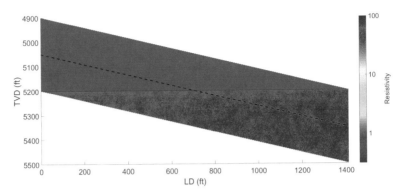

Fig. 5.8 SDAR A3 model. Two-layer landing scenario. The resistivity value of the upper conductive layer is 1 ohm-m and the lower layer is 50 ohm-m. The tool is drilling at the dip angle of 78° along the dash line

before. The initial state is set to a homogeneous conductive model. The resistivity is 1 ohm-m. In a parallel tempering scheme, 16 samplers are launched simultaneously where the temperature values exponentially increase from 1 to 1000.

Another indicator to examine the performance of this method is the processing efficiency. To satisfy the requirement of real-time formation evaluation, the time cost for the sampling process in a single logging station should be limited in minute-scale. Given the total chain length, which is identical to the number of evaluations of forward function, the time cost is around the run-time of a single chain on a single processing unit. It should be known there is an overhead in the communication during the chains swap. Hence, there is a trade-off between the times of swaps and parallel efficiency. In our experiment, the length between the two swaps is set to about 0.1% of total chain length. Except for the chains swap, all chains run in parallel independently. All the experiments are simulated on a desktop with a multi-thread Core i7 CPU. A set of Standardization of Deep Azimuthal Resistivity (SDAR) benchmark models are used.

We start the example from the SDAR model A3, which is a two-layer landing case. Figure 5.8 depicts a high-angle landing job, where the borehole has 78° relative dip angle respecting to the bed. Starting from 5050 ft in TVD, the formation boundary is 150 ft below the measurement point, and the drilling tool proceeds with measuring at every 2 ft in TVD. The resistivity of the upper conductive layer is 1 ohm-m, while the lower resistive layer is 50 ohm-m. The drilling trajectory stops at 5350 ft in TVD, and 151 logging stations are created. To maximize the detection range for the landing purpose, the investigating scope is set to ±150 ft as from tool depth in this case. The maximum chain length is set to 8k. We compare the results between single-chain tMCMC and multiple chains with parallel tempering in Fig. 5.9.

We use the mean value of the last 2k samples as the inversion results at each logging station. 151 individual inversion results group the 2D subsurface image. The average time spending for each point is about 35 s for the single chain tMCMC and

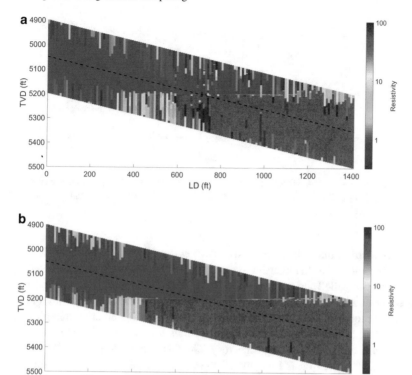

Fig. 5.9 The inversion results of SDAR A3 model with 151 logging points. The sampling of 1D model is conducted at each logging point. The mean value of the last 2k samples is used as the inversion result and grouped to construct the 2D image. Result by using parallel tempered tMCMC has better quality than using a single chain tMCMC with the same number of iterations

42 s for tMCMC with 16 parallel chains. The overall imaging quality is significantly improved with parallel tempering. In general, the result demonstrates a basic two-layer model with one primary interface. The poor quality of Fig. 5.9a results from the insufficient sampling; in other words, the chain is not long enough to explore the PPD. Parallel tempering takes advantage of the parallel computing system and boosts the sampling efficiency that recovers the model structure from the noisy data more accurately with limited iterations. We use a trace plot to demonstrate the behavior when chains exchange states and gain more dynamics in the sampling.

In Fig. 5.10, it shows the convergence of Markov chains at the logging station in 5220 ft in TVD. The data misfit value in logarithm scale is used to show the change during iterations. The red curve denotes the cold chain that samples the target distribution while the gray chains in the background are sampling the tempered distribution. An obvious leap was detected around the 400th iteration, which reduces the data misfit significantly. This is a successful swap of states between the cold chain and one of the hot chains. The state in the hot chain introduces another region of the

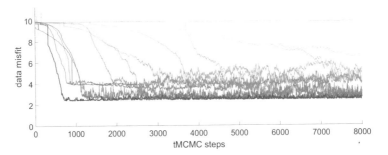

Fig. 5.10 The trace plot of Markov chains of the logging point at 5220 ft in TVD. The convergence of Markov chain is boosted by parallel tempering, in which the cold chain exchanges information with other hot chains

sampled distribution to the cold chain and makes it gain the dynamic and quickly jump to another model in this distribution and do the sampling. Usually, the converging iterations are called the burn-in period. The parallel tempering boosts the burn-in speed. For a real-world application, this process can be monitored through the trace plot and autocorrelation plot. The sampling can be terminated once enough samples are yielded. In this example, the interpreter tells a two-layer simple model for the most points along the drilling trajectory. The frame combining parallel tempering with tMCMC gives clear imaging of reconstruction of a two-layer earth section profile.

The second case uses the SDAR B1 model, which is shown in Fig. 4.6. The reconstructed image of a 2D earth profile suffers from a little bit high noise with a single chain. Moreover, the challenge during the interpretation process comes from the first approaching thin bed with 5 ft thickness. It is not always easy to have this structure inverted especially when the typical prior information has relatively low resolution. In the model-based inversion, the number of model layers should be chosen carefully. With the same setting as the first example, we conducted independent 1D statistical inversion along the trajectory from 4930 ft to 5150 ft in TVD at 1-foot step. There are 221 logging stations involved. The apparent dip angle between wellbore and bed is 78°. We apply the same level of noise to the synthetic dataset. The maximum chain length is 25k with the last 5k samples yielded to infer the statistical results. We compare the overall performance by launching a single chain tMCMC with parallel tempered tMCMC.

Figure 5.11a is the result using single-chain tMCMC method, which has been shown in Chap. 4. The mean value of last 5k samples is used as the inversion result and combined to this 2D figure. The poor image quality still results from the limited sampling given 25k chain length. The single point evaluation costs about 2.3 min, whereas the inference of model complexity and the statistics of model parameters has not converged to the desired outcome, although the basic model structure can be approximated. Figure 5.11b is the enhanced result by running 16-chain parallel tempered tMCMC. The imaging quality is refined greatly while the average inverse

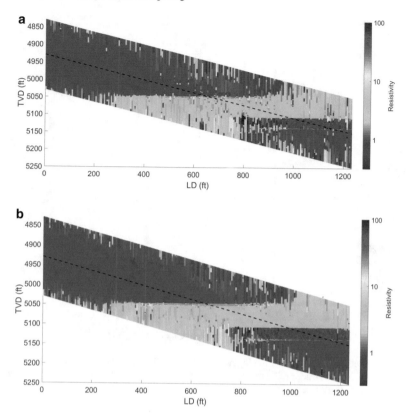

Fig. 5.11 The inversion results of SDAR B1 model with 221 logging points. A five-layer structure can be resolved even a strong noise is applied to the measurements. Parallel tempered tMCMC still returns a better imaging quality than a single chain tMCMC

time of a single point is near 3 min, taking a full-length iteration. The image noise comes from the measurements error inherently. However, the interpretation is successful where an apparent thin bed is resolved smoothly and continuously when the tool is around 50 ft above it. The distance between the thin bed and the main target is about 56 ft vertically. It is noted that the image of the main target (the second resistive layer) is resolved poorly until the tool crossed the thin bed. We take further investigation at the logging stations in the middle conductive layer.

Figure 5.12 shows the statistics at the logging depth of 5060 ft in TVD. The peaks of interface histogram present in multiple depths, which denotes a potential multi-layer structure. Four different regions can be distinguished when looking into the resistivity probability map on the right panel. Starting from the shallower depth, an apparent conductive region can be seen, though some small artificial interfaces exist, which may result from the measurements error propagating to the parameter space. The second small region distinguished from neighboring layers is the 5 ft thin bed that has been well resolved with two clear boundaries at 5050 ft and 5055 ft in

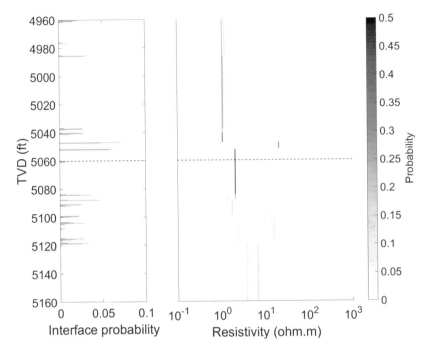

Fig. 5.12 The statistical result at the logging depth of 5060 ft in TVD, where the tool is near the thin bed. Multi-layer structure is presented, but the structure below the tool is poorly interpreted with large uncertainty

TVD. The third region is the current formation where the tool locates in. Owning to the measurements that sensitive to the local resistivity, the resistivity in current formation is often inverted preciously with minimal uncertainty. The uncertainty of resistivity in the lower layers is enormous. The interpreter returns multiple modes to the resistivity value given a certain depth. It is explained that the inverse problem is ill-posed; the distribution of model parameters is multi-modal. Hence, the model structure and the value of earth parameters are uncertain.

We investigate the statistics when the tool is moving forward and approaching to the main target. Figure 5.13, as a comparison to Fig. 5.12, shows the statistics at the logging depth of 5088 ft in TVD where the tool moves to the middle part of the conductive layer. From both interface probability and resistivity probability map, a five-layer structure is clearly presented. Though the resistivity value and the interface depths of the thin-bed gain more uncertainty, the lower two layers, particularly the upper and lower boundaries of the resistive layer are resolved well. The resistivity value of the deepest formation has more substantial uncertainty with wider color span. However, the interpretation at this depth is very successful.

At last, we show the trace plot of Markov chains at this logging station in Fig. 5.14. Two major swaps between the cold chain and the hot chain can be seen at around 2.5k

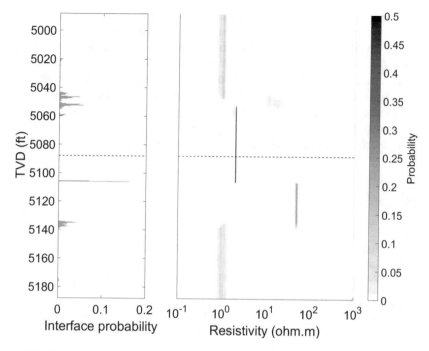

Fig. 5.13 The statistical result at the logging depth of 5088 ft in TVD, where the tool is in the middle of conductive formation. A five-layer structure is clearly interpreted

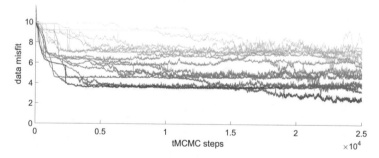

Fig. 5.14 The trace plot of Markov chains of the logging point at 5088 ft in TVD. Two swaps between the cold chain and the hot chain are detected, which leads the convergence. The last 5k samples are valid to examine the statistic of model properties

and 18k iterations, which lead the cold chain to converge from the burn-in period. The last 5k samples are yielded for the analysis of the sample statistic.

In this chapter, we witness that the tempered Bayesian samplers can interpret the complex measurements that influenced by the multi-layer earth structure. The resolution to the thin bed is satisfactory through the image of the final result. For a complicated model inversion using tMCMC sampling, the overall sampling quality is significantly influenced by the chain length or the total number of samples. With limited iterations, we see that parallel tempering can be a powerful technique to boost sampling performance and help achieve both quality and efficiency at the same time.

References

1. Schafer JL (1999) Statistical methods in medical research 8(1):3
2. Shen Q, Wu X, Chen J, Han Z (2017) Appl Comput Electromag Soc J 32:5
3. Lu H, Shen Q, Chen J, Wu X, Fu X (2019) J Petrol Sci Eng 174:189
4. Neiswanger W, Wang C, Xing E (2013) Asymptotically exact, embarrassingly parallel MCMC. arXiv:1311.4780 (2013)
5. Gelman A, Rubin DB (1992) Statist Sci 7(4):457
6. Aarts E, Korst J (1988) Simulated annealing and Boltzmann machines. Wiley, New York, NY
7. Geyer CJ (1991) Markov chain Monte Carlo maximum likelihood. Interf Found North America

Index

Printed in the United States
By Bookmasters